GENETIC MARKERS OF SEX DIFFERENTIATION

REPRODUCTIVE BIOLOGY

Series Editor: Sheldon J. Segal
The Rockefeller Foundation
New York, New York

THE ANTIPROGESTIN STEROID RU 486 AND HUMAN FERTILITY CONTROL
Edited by Etienne-Emile Baulieu and Sheldon J. Segal

CONTRACEPTIVE STEROIDS: Pharmacology and Safety
Edited by A. T. Gregoire and Richard P. Blye

GENETIC MARKERS OF SEX DIFFERENTIATION
Edited by Florence P. Haseltine, Michael E. McClure, and Ellen H. Goldberg

GOSSYPOL: A Potential Contraceptive for Men
Edited by Sheldon J. Segal

IMMUNOLOGICAL APPROACHES TO CONTRACEPTION AND PROMOTION OF FERTILITY
Edited by G. P. Talwar

A Continuation Order Plan is available for this series. A continuation order will bring delivery of each new volume immediately upon publication. Volumes are billed only upon actual shipment. For further information please contact the publisher.

GENETIC MARKERS OF SEX DIFFERENTIATION

Edited by

Florence P. Haseltine and
Michael E. McClure
National Institute of Child Health and Human Development
Bethesda, Maryland

and

Ellen H. Goldberg
The University of New Mexico
Albuquerque, New Mexico

PLENUM PRESS • NEW YORK AND LONDON

Library of Congress Cataloging in Publication Data

Workshop on Genetic Markers of Sex Differentiation (1986: Center for Population Research)
Genetic markers of sex differentiation.

(Reproductive biology)
"Proceedings of the Workshop on Genetic Markers of Sex Differentiation, sponsored by the National Institute of Child Health and Human Development, held October 9–10, 1986, at the Center for Population Research of the National Institute of Child Health and Human Development in Bethesda, Maryland"—T.p. verso.
Includes bibliographies and index.
1. Sex differentiation—Congresses. 2. Sex determination, Genetic—Congresses. 3. Immunogenetics—Congresses. I. Haseltine, Florence. II. McClure, Michael E. III. Goldberg, Ellen H. IV. National Institute of Child Health and Human Development (U.S.) V. Title. VI. Series. [DNLM: 1. Genetic Marker—congresses. 2. H–Y Antigen—congresses. 3. Sex Differentiation—congresses. QS 640 W926g 1986]
QP278.W67 1986 612'.6 87-25490
ISBN 0-306-42679-X

Proceedings of the Workshop on Genetic Markers of Sex Differentiation, sponsored by the National Institute of Child Health and Human Development, held October 9–10, 1986, at the Center for Population Research of the National Institute of Child Health and Human Development in Bethesda, Maryland

A Division of Plenum Publishing Corporation
233 Spring Street, New York, N.Y. 10013

Printed in the United States of America

To my father, William R. Haseltine, Ph.D., who answered my never-ending questions about the two-sex problem until, filled with exasperation, he said, "When you grow up, you figure it out!"

FPH

To the late Professor Lubomir Sidonius Hnilica, Ph.D. (January 26, 1929–March 30, 1986), Stahlman Professor of Cancer Research, mentor, colleague, and friend, whose shining example guides my footsteps.

MEM

To Gertrude Heidenthal, Ph.D., who sparked my interest in genetics and to whom I shall always be grateful.

EHG

CONTENTS

INTRODUCTION - NATIONAL INSTITUTE OF CHILD HEALTH AND HUMAN DEVELOPMENT

(Mechanisms of Sex Differentation)

Florence P. Haseltine

A workshop on the Mechanisms of Sex Differentiation was held at the Center for Population Research, at the National Institutes of Health, in October 1986. The purpose of this meeting was to provide a forum for leading scientists in the field of sex determination research to discuss their most recent results. The organization of the meeting permitted extensive discussions of evolving ideas, technologies, and biological models associated with clinically relevant molecular mechanisms. It is anticipated that productive new concepts and research directions will have emerged from this meeting.

The Center for Population Research (CPR) of the National Institute of Child Health and Human Development (NICHD), which sponsored the workshop, sponsors programs charged with promoting research in reproductive biology and medicine which will foster technological advances and lead to the improved reproductive health of the human population.

The mission of the CPR includes the elucidation of gonadal function for fertility regulation. The interaction of the germ cell and stroma is central to our mission and so is their development. The areas of interest to us, therefore, include the events surrounding early sex differentiation that affect: 1) gametogenesis; 2) gonadal differentiation; and 3) early genetic expression of female and male specific proteins which infuence these events.

Sex differentiation starts at the moment of fertilization and the male and female mouse embryo show differences that can be detected by the 8-cell stage. The genetic material in both sexes is constitutively different from that stage to adulthood. These differences occur well before implantation and have dramatic implications for gametogenesis and gonadogenesis.

Disorders of sexual differentiation represent a significant, unresolved, clinical problem of the human condition and are associated with two of the most basic aspects of human life--gender identity and the ability to reproduce. The research program of the Reproductive Sciences Branch of the NICHD seeks to advance knowledge of the molecular basis of disorders of sexual differentiation and their reproductive consequences. The differentiation process is driven by the presence or absence of the Y-sex chromosomal material in an individual's tissues. Although molecular mechanisms of selective gene expression must control these processes, these mechanisms remain poorly understood at present. Since 1984, exciting new results have come from the use of the techniques of molecular biology to study sex determination. These molecular studies have uncovered facts which may prove to be of profound clinical importance. New findings and the extraordinary potential promise of genetic markers for sex determination require that contemporary scientists reassess the model(s) of sex determination. This meeting provided a forum for the reassessment of these models, by introducing the new results and providing the investigators with an opportunity to consider and discuss them together.

The editors of this volume have been actively involved in research into the mechanisms of sex differentiation from different perspectives. The stimulus for the meeting came from the prior involvement of the editors and authors in the field of sex differentiation and their commitment to the idea that problems of sex differentiation can be greatly clarified by the introduction of the techniques and principles of modern molecular biology.

Many people have invested considerable effort in the study of sex differentiation. This volume represents the work of only a small fraction of those contributors. Meetings of the present type are, naturally, limited with respect to the number of active participants. Nevertheless, all the people who attended, whether they spoke, wrote, asked questions or simply listened, contributed to the goal of the workshop. The introduction which follows discusses the work that was presented and, since the field is so vast, no disrespect is meant to those contributors who are not cited. The contributors to this volume are all directly involved in various areas of sex differentiation.

Sexual reproduction is the common form of reproduction in the animal kingdom, for reasons which have been the basis of speculation by scientists for as long as there have been scientific records. Such speculation began when it was realized that sexual reproduction is necessary to embed the successful element of genetic diversity in the next generation of offspring.

Both intracellular and intercellular mechanisms of development must be studied in order to examine adequately the many models of sex differentiation. Although sex differentiation is the expression of pairing of homozygotic and heterozygotic chromosomes, the somatic expression of this difference is not completely understood. The study of the translation of genetic sex into phenotypic sex has taken several forms, each of which has tested the ingenuity of researchers in the fields that it has touched. The longer that developmental differentiation is studied, the more detailed the questions become. Every new technological breakthrough in the field of molecular biology has lead to new and immediate applications to problems in sex differentiation. The techniques which have contributed important information to our understanding of the mechanism of sex differentiation came from the fields of classical genetics, immunology and molecular genetics. Researchers in these three fields have interacted beautifully, as the results of collaborations presented at this meeting demonstrate.

Differentiation is initiated at fertilization by the fusion of an egg and a sperm, each of which carries a sex chromosome. The translation of the genetic sex of the fertilized egg into phenotypic sex was one of the topics of this symposium. Methods for evaluating the regulation of gene expression and the interactions of the cell surface commponents were presented. Examples were given and discussed of environmental effects on sex differentiation.

One of the reasons for evaluating several different animal species is that the small differences in the control of differentiation between the various systems permit the elucidation of the process as a whole. Sex differentiation in reptiles, eutherian mammals and marsupials was discussed in detail. A generalized observation arises from such studies: two sexes are required for reproduction of a species, because deviation from that norm limits further evolution of that species, i.e., permutations away from that system produce an evolutionary dead end.

Every field needs a theoretical focus. While many people have wondered why there are two sexes, few of the leaders in the theoretical approach to the evaluation of sex differentiation have had the conceptual impact of Susumo Ohno. He is fascinated by the antiquity of the mechanism of sex determination and parallels his theories of patterns of sex differentiation with many other examples from nature of preserved patterns. In developing his theories, Ohno examines the cases where differentiation has been derailed. He struggles with the situations of sex reversal which have arisen in the abnormal situation represented by a mosaic individual, and in the adaptive situation represented by the reptile in which sex differentiation is dependent on external environmental stimuli. Perhaps the most interesting contribution to the meeting was his musical tribute to sex differentiation. To appreciate his discussion of the evolution of the triplet code expressed in gene sequences, it is necessary to examine his models as they are presented in this volume. For the participants at the meeting, the development of his idea of correlating a repetitive coding pattern with underlying rhythms culminated in his putting the triplet code for the nucleotide sequences of certain genes to music. Some compositions were as pleasant to the ear as portions of fine, classical symphonies; some were as jarring to the auditory senses as some dissonant modern compositions. Ohno has contributed to the field as both a theorist and an experimentalist. In this volume, he presents his theories on the influence that sexual reproduction has exerted on the evolutionary conservation of certain nucleotide sequences in DNA.

Other speakers also presented evidence that there is more than one way to interpret data and formulate theories. Roger Short has been credited with the hypothesis that female sex differentiation is passive, but it is hard to consider the female genital tract as a default condition. As a social statement it is certainly not valid, and from a biological perspective there is no differentiation that can be considered passive. Nonetheless, he and his collaborators are making a very careful evaluation of the control of sex differentiation in the fascinating marsupial model.

Short presented an innovative study of the development of marsupials and demonstrated how their differentiation presents special opportunities for the examination of factors which control sex differentiation. The young marsupial offspring is difficult to sex at birth. Both the gonadal and phenotypic expression of sex develop after birth. These animals provide an opportunity for the evaluation of the possible role of hormonal effects on sex differentiation and the study of the effects of other environmental situations.

David Crews presented some of his work on reptiles and birds. Birds are the mirror image of mammals in terms of expression of heterochromatic sex differentiation. Reptiles have contributed considerably to our understanding of sexual differentiation because their sexual differentiation can be determined by common external environmental influences, namely the fluctuations in temperature which occur during the incubation of the egg. The demonstration that the central nervous system is developmentally sensitive to temperature was intriguing. Even in lizards where parthenogenetic females exist, the pairing of two females elicits an alternating female- and male-like behavior which is necessary for the production of more parthenogenetic females. So even here, where no gametes are physically exchanged, mating behavior is needed.

Ernst Eichwald presented a contemporary perspective on his original observation that there is a consistent difference between males and females with respect to certain antigens found on the surface of transplanted tissues. Although he credits serendipity with the selection of the correct specific strain of mice for the continuation of his transplantation experiments, it took both painstaking observations by his technician and the careful examination of data to reveal that the rejection of skin in his experiments was specific only to male skin transplanted to females. The significance of the data could have easily been overlooked or ignored because there was then no reliable way of interpreting experiments on rejection of skin since, at best, the experiments were successful only 50% of the time. This antigen was originally considered to be a minor Histocompatablity Transplantation Antigen but later became known as the H-Y antigen.

Eichwald's landmark experiment, and the many that followed which showed that this antigen was ubiquitious, produced much speculation and many theoretical constructs. Controversies even about the name of this substance and whether or not it is a single product have filled the literature. Much work has been done in an attempt to elucidate the genetic control, function, and various forms of this H-Y antigen.

The H-Y antigen has at one time or another been credited with many wonderful charms and attributes. In fact, so many functions have been assigned to H-Y antigen that there is no way it can live up to its reputation. The properties which were attributed to the antigen derived initially from the prediction that it was the substance that promotes testicular differentiation. Since there are examples of testicular differentiation without the concommitant appearance of male germ cells, it was predicted that H-Y caused the production of testes. H-Y antigen was considered to be diffusible and transferable between cells, and its action could be blocked by either the germ cells or a gonadal organizing substance of the opposite sex. It was also held that H-Y antigen was recognized by a specific receptor and that this receptor should be found within or on the target cells which respond to testicular or ovarian organizing signals. However, it has recently become evident that the testicular organizing locus on the Y chromosome is separate from the locus of the H-Y antigen. Obviously, the definition of the H-Y antigen has undergone a remarkable expansion from that in the original hypothesis.

Gloria Koo has reviewed the history of H-Y antigen studies with a detailed discussion of the serologically detected H-Y antigen and has discussed how the antigen may be involved in sex differentiation. Her chapter includes the methodology used to evaluate properties of the antigen. One of the difficulties with any of the assays for the antigen is that the antibodies raised have always been of low titre. Even monoclonal antibodies have not been obtained with high titre activity.

Indeed, the antiserum has been notorious for giving results of poor reproducibility. Its use has required experienced personnel, and it has a cytotoxic capability which is often sensitive to interaction with complement. The assays have been difficult to reproduce and to validate, and although several laboratories have had assays for H-Y at one time or another, it has been a system that can be employed only sporadically. Setting up of the assay from scratch after a hiatus has always been very difficult. Eichwald pointed out that there is no laboratory in this country that can be routinely called upon to run the assay. Furthermore, there may be only one chance to perform an assay on a given specimen, and there is frequently only a small window in time during which the sample can be obtained, so that the logistics of running such an assay are very complex.

Koo argues that the serologically detectable antigen and the antigen detected by cellular assays are similar. Her studies with blocking antibodies, the development of new assays for H-Y antigen with ELISA reagents and monoclonals have been extremely important in the study of H-Y antigen, and her work may one day permit the testing for this family of antigens on a widespread basis.

Serologically positive H-Y antigen has been found in patients with abnormal X chromosomes and detectable antigen has been measured in patients who have a mosaic karyotype. In a related discussion, Anne McLaren has summarized the arguments about Turner's syndrome mice and the H-Y antigen.

Many of the patients under study whose serotype is positive for H-Y antigen are also mosaic for a line of cells that is 45,X with either a Y chromosome or another chromosomal abnormality present in another cell line. Initially it was noted that most of these patients had serologically detectable levels of H-Y antigen which were no different from those of normal male controls, although intermediate levels of activity may be present.

Frequently as few as 3% of cells in mosaic patients have the mosaic karyotype. These circumstances suggest that while levels of serologically detectable H-Y antigen are regulated by the Y chromosome, the actual cell surface or biological effect of the H-Y antigen is regulated by another chromosome which no longer requires the initial regulation by Y chromosomal material. The expression of H-Y antigen on the cell surface is known to occur very early in development. Ellen Goldberg's early work on expression of serologically detectable H-Y antigen demonstrated its presence on the cell surface of mouse embryos at the eight-cell stage. The implication is that once a chromosome (an X or an autosome) in a given cell has started to express H-Y antigen, all of the descendants of that cell will continue to express H-Y antigen if differentiation is maintained.

McLaren's presentation represented a bridge between those researchers who study sex differentation at the physiological level and those who have tried to make some sense out of the role of H-Y antigen from a genetic viewpoint. Her genetic experiments and analyses of the research on the chromosomal role in mice have done much to clarify the function of the Y chromosome in sex differention. Sex-reversed mice, namely those mice with XX chromosomes which have testes, have now been demonstrated to have Y-chromosomal material translocated to the X chromosome. This genetic material originates from the distal end of the short arm of the Y chromosome. This region pairs with the pseudoautosomal segment of the X chromosome and, in some situations, will produce a translocation which carries the testis-determining gene with it. How much of the Y chromosome is translocated determines whether regulator for H-Y antigen is translocated with the testis-determining factor or not. There is, however, no production of sperm without the Y chromosome and, thus, it seems that other chromosomal segments which are not translocated are necessary for production of spermatogonia.

McLaren's work with both the transplantation form and the serologically detected form of H-Y antigen shows that sex-reversed mice may or may not carry the translocated H-Y locus or loci, and, therefore, the testis-determining gene is separate from and independent of the H-Y locus.

More work is required to identify the genes which are involved in spermatogenesis and to determine whether they are linked to or are identical to any of the forms of H-Y antigen. The possibility that an antigen causes cellular differentiation has been given new life by the finding that T cells in the lymphatic system respond to antigens on other cells by induced proliferation. Whether a similar role can be found for an antigen-receptor complex on the spermatogonium awaits further experimentation. McLaren's experiments on mosaic mice indicate that when XY germ cells are in an XX ovarian environment they develop as oocytes. Her observations are critical for an understanding of the interactions between the sperm cells and the somatic cells.

Claude Nagamine describes hermaphrodites and examines the genetics of their sex differentiation. His work, coupled with that of other investigators, implicates the steroid sulphatase locus, on the X chromosome near the centromeric region, as a marker of an activity which is involved in sex differentiation.

Peter Goodfellow presented his work on the Y chromosome, and noted that only 7 genes have been mapped to the Y chromosome, and only one of these has been cloned and sequenced. He has used pseudoautosomal genes of the Y chromosome to calculate crossover frequencies and genetic mapping distances. This approach has permitted a mapping of the distal end of the Y chromosome. There are very few genes on this chromosome and they appear to fall into only two classes. Because the Y chromosome performs a most important function, it has been hypothesized that it has been spared the usual housekeeping chores. Only critical functions appear to be preserved on this chromosome, and no functions have been permitted to remain which detract from the value and functioning of this early determinator activity. Goodfellow's discussions on the similarities and differences between the mouse genetic data and the human genetic data highlight the difficulties found in analyses of genetic maps of the Y chromosome.

Different classes of DNA sequences provide important markers on the Y chromosomes. The difficulty which results from looking at only one model was further discussed by Kirby Smith. He compared the Y chromosomes of many related species. It is of particular note that the Y chromosome is different in related species. In humans, there is much more repetitive DNA than in related primate species. The reasons the human Y chromosome has collected repetitive DNA, while related species have not, may yet prove to be significant. The structure of the human Y chromosome is distinctly unique. Four classes of repeat sequences have been identified, and Smith has studied these in depth with Y-specific and non-specific probes. He has studied, in particular, the 2.1 and 3.4 kilobase repetitive sequences which are found on the Y chromosome and has found a clustering of repetitive sequences on the long arm. There are polymorphisms among these repeated sequences and they are not the same as the repeating sequences found on autosomes. The Y chromosome is dynamically conserved with a high degree of rearrangement within a species. Nevertheless, there is only a small degree of conservation between species. Most of the sequences on the human short arm are homologous to sequences on the X chromosome but do not have any homology to sequences on the Y chromosomes of other primates. The origin of at least one half of the Y chromosome remains totally unclear. Differences between man and the primates seem to have evolved through a series of translocations, rearrangements and mutations, with a large degree of potential evolution which bears further examination.

David Page, Albert de la Chapelle and Kirby Smith have worked together and in their own laboratories to map the Y chromosome. Page and de la Chapelle both presented recent work on the mapping of the testicular differentiation locus in the human, based on hybridization studies with cloned Y-DNA which permitted deletion mapping. Page has also isolated additional clones of DNA from the Y chromosome by "chromosome walking". These clones are being studied to determine the occurrence of open reading frames and certain of these clones may serve to retrieve mRNA for further genetic analysis. Page has mapped the testis-determining gene to an interval on the short arm of the Y chromosome. By deletion mapping of the Y chromosome material (not including the pseudoautosomal regions), patterns have been identified which suggest that both XX males and XY females arise from translocations of the 1A segment of the Y chromosome. The testis-determining locus is just below the pseudoautosomal region of the Y chromosome. The locus for the serologically detected H-Y antigen is separate from the testis-determining locus and maps to 4B (the centromeric region) and not to 1A. Page concludes that the H-Y locus cannot be involved in testis determination but does not rule out that it may be involved in the regulation of spermatogenesis. Each XX male and each XY female appear to result from a new translocation and not as the result of a familial inherited trait. However, it remains possible that what might be heritable is a fragility in this region of the chromosome, which is distinct from any other recombination-related, fragile chromosome syndrome. The mapping employed a Y chromosome marker called PPD 230, a marker specific for the X and Y telomeric regions. This region has been demonstrated to be involved in the translocation of material from the father's Y chromosome to his X chromosome, which then gives rise to the XX sex-reversed offspring.

Robert Erickson has examined the role of the small, mapped nucleotide sequences, called Bkm sequences, with regard to their possible role in sexual differentiation. These repeated sequences are so common that a key role has been sought for them, especially since they appear to be conserved, male-specific sequences, as detected by restriction enzyme analyses. While the Bkm sequences are located on both the sex chromosomes and autosomes, with a concentration of these sequences on the sex chromosomes in the mouse, this distribution does not appear to hold in the case of humans. It does appear, however, that the Bkm-related sequences on autosomes are conserved, whereas the sequences on the Y chromosomes are not. Erickson concludes that there is presently no evidence to assign a major role in sex differentiation to these Bkm sequences.

Jonathan Wolfe presented work on the one set of repeat sequences on the Y chromosome. There is a group of nucleotide sequences, called alphoid sequences, which are tandem repeats of a particular subunit. Humans have been found to have more repeated sequences than do other primates. Alphoid sequences occur in the centromeric region of chromosomes and do not appear to be transcribed. The repeats of alphoid DNA, and the divergences within these regions, may only be by-products of past amplification processes which occurred during gene recombinations. The alphoid repeats found on the human Y chromosomes provide a source of these unusual DNA sequences and their abundance provides a unique opportunity for the evaluation of the structural attributes of these sequences and the elucidation of their function. H-Y antigen regulation and spermatogenesis have also been mapped to this region. The concentration of these sequences around the centromere makes them contenders for participation in meiotic and mitotic functions. H-Y antigen regulation and spermatogenesis are mapped to the centromeric region. The concentration of these alphoid sequences of DNA around the centromere of the Y chromosome, a segment required for spermatogenesis, permits speculation as to their role in spermatogenesis and possible expression of H-Y antigen. It will be interesting if the regulator sequence for H-Y antigen is found in this region, even if the protein which processes H-Y antigen is encoded on another chromosome.

Yun-Fai Lau's presentation on a male enhanced antigen is work in its early phases. A cDNA library was made from testis and was incorporated into a bacterophage with a viral expression vector. The phage plaques which were identified by polyclonal H-Y antiserum, present a dilemma. The proteins that may be related to H-Y antigen could share glycosylated structures and may not have been identified. This method of cloning could also have cloned DNA for another protein that is male specific. Careful controls and more specific antiserum will be needed to make sure the product is male-specific, and it will be some time before its identity and putative relationship to an H-Y antigen is clarified.

Paul McDonough presented results from his laboratory on the detection of DNA which originated from the Y chromosome in intersex subjects where no Y chromosomal material could be found by standard karyotypic analysis. This work has a great deal of clinical relevance. Since it is not yet possible to identify all those patients with Y chromosomal material by cytogenetic techniques, McDonough has used the probes developed by Smith, de la Chappelle, and Page to find fragments of Y chromosomal material in the DNA of patients for whom other tests could not clarify chromosomal abnormalities. As McDonough points out, individuals with Y chromosomal material are sometimes susceptible to the development of dysgenetic tumors. It is not yet known what properties of the Y chromosome or its DNA sequences are involved in these malignant tendencies. If the genetic contribution of the Y chromosomal material is to be stable, there must either be centromeric material present or the DNA fragment must be translocated to a different chromosome. It would be interesting to determine whether these Y chromosomal fragments are protooncogenes by themselves, or whether they activate a locus when placed on an inappropriate background, for example in the presence of 2 X chromosomes. The expression of the malignant potential is also known to be hormonally influenced. McDonough cautions that the probes he has studied so far are specific for the long arm of the Y chromosome. Nothing has been mapped to these probes, so further studies are clearly still required.

The participants in the Workshop heard new results and hypotheses, discussed them in the light of previous data, and left with renewed enthusiasm for future experiments. It is the purpose of this volume to convey some of this enthusiasm to a wider audience than could be accommodated in one room, and to act as a catalyst in the design of new experiments to shed still more light on the various mechanisms of sex determination.

Our thanks are extended to Ms. Penny Colbert for typing and retyping the manuscripts. She made sure everything ran well behind the scenes. Dr. McClure paid attention to more details than can be enumerated. Both Drs. McDonough and Goldberg found excellent speakers in their respective areas of expertise. Dr. Ann Korner served as our relentless manuscript editor.

Finally, I would like to extend my personal thanks to the National Institute of Child Health and Human Development.

EVOLUTIONARY INSIGHTS FROM REPTILIAN SEXUAL DIFFERENTIATION

David Crews and James J. Bull

Department of Zoology
University of Texas at Austin
Austin, Texas 78712

MODEL SYSTEMS AS GENERALITIES

A major objective of any scientific endeavor is the discovery of generalizations around which various observations may be organized. In the many physiological disciplines of biology, generalizations have often been sought in the form of a model system, i.e. a physiology determined from studies of one or a few species that is projected as a possible model for other taxa. Two such major generalities are the genetic code and Krebs cycle of respiration. Model systems are of immense value in providing a focus for studies from a diversity of species. However, whenever a model system is based on results from a group of closely related taxa and a new group is then studied, one must be prepared to accept alternative models. Principles are limited by the exceptions to them.

The discovery that a single model system fails to generalize to a group of species leaves two avenues open for the further pursuit of generality. First, the diversity observed at one level of analysis may suggest other levels of analysis where model systems do apply. Second, generalities may be discovered in the explanation of diversity itself. For example, when taxa are found to vary, an evolutionary perspective may be employed to facilitate understanding the variety. By understanding what selective pressures have been present to change certain systems, as well as by observing the variety that has actually evolved, both the variety and stasis among species may be identified and explained.

The purpose of this paper is to use some of our work on sex determination and sexual differentiation to show both how and where model systems derived from mammals and birds fail to serve as the basis for generalizations about reptiles (as addressed also by Licht in 1). We will consider three topics: (i) sex determining mechanisms; (ii) the role of gender in the development of sexually dimorphic behavior; and (iii) the influence of steroid hormones on the organization of neural circuits subserving sexual behavior. In addition to illustrating how reptiles challenge current paradigms based on model systems, we suggest ways in which an evolutionary perspective may offer insights that increase our understanding of the diversity, and we suggest that generalities may yet be found at other levels of physiology. We close by pointing out some cases where natural variation in mating systems offers promise for the integration of physiology and evolutionary studies.

SEX DETERMINING MECHANISMS

As recently as 1967, the established tenets of vertebrate sex determination were that all amniotes and amphibians have male or female heterogamety, although groups differ considerably in the magnitude of sex chromosome differentiation that accompanies the heterogamety[2,3]. Ohno[3] has presented, in fine detail, the different degrees of X-Y heteromorphism among vertebrates, from strict homomorphism in frogs and fish, to intermediate stages in some snakes, to extreme heteromorphism in mammals, birds, and yet other snakes. Although sex chromosome heteromorphism was not ubiquitous, heterogamety occured in all groups, suggesting that it was older than the origin of amniotes.

Just when this orthodoxy seemed unquestionable, Charnier[4] discovered that the temperature of incubation controlled the sex of hatchlings in the lizard *Agama agama*. His discovery went largely unnoticed, but in the early 1970's, Pieau[5] began to report the same phenomenon in two species of European turtles. By 1980, it was clear that *temperature-dependent sex determination* (TSD) was a genuine mechanism that operates in many populations of reptiles[6]. Heterogamety is, therefore, not ubiquitous in amniotes, and there is, in fact, no reason to suggest that heterogamety was even ancestral to amniotes. In the next few paragraphs, we review the basic information currently available about TSD.

A variety of environmental variables has since been found to influence the sex ratio in vertebrates, including temperature, soil water potential and pH. Incubation temperature has strong effects on sex ratios of hatchlings in many turtles, a few lizards, and perhaps all crocodilians (discovered in lizards by Charnier, 4; in turtles by Pieau, 5; in crocodilians by Ferguson and Joanen, 7; reviewed in 6, 8). Under TSD, any embryo can develop as male or female depending on incubation temperature, and major inherited effects on sex determination are precluded. Sex chromosomes have not been observed in any of the many species known to have TSD.

Reptiles with TSD exhibit various relationships between temperature and sex ratio. Low temperatures produce all females and high temperatures produce all males in the cases of lizards and alligators, whereas this pattern is reversed in turtles[8]. This effect is not due to differential mortality, and is permanent, extending to adulthood[9]. The temperature-sensitive period of sex determination occurs in the middle of development, coinciding with gonadal differentiation in crocodilians, turtles, and lizards with TSD (reviewed in Bull, in press). The sex of the embryo depends on a cumulative effect of temperature during this period, rather than on the temperature in a narrow window of development.

While the majority of work on TSD has been conducted in a laboratory setting, several field studies have confirmed its operation under natural conditions. Depending on the species, the sex ratio of offspring in individual nests has been shown to vary according to sun/shade exposure, depth in the soil, elevation above water, and season of deposition[11-14]. Manipulation of eggs in the natural environment has further shown that the position of the nest influences sex determination (rather than a predetermination of sex by the maternal choice of an appropriate nest site). Records of daily temperatures in several natural nests of turtles reveal strong associations between temperature and sex ratio, agreeing quantitatively with laboratory data in one study[15] though not in another[16].

An evolutionary perspective may suggest why heterogamety is not ubiquitous and why the original paradigm is invalid. The possible evolutionary significance of TSD over genotypic sex determination (GSD) has been considered in a general context by Charnov and Bull[17] and applied to the reptile case by Bull[6,8]. Briefly, models indicate that TSD will evolve in place of GSD if incubation temperature influences an individual's fitness as male versus female. Thus, if an animal developing at warm temperatures would, years later, experience higher fitness as a male than as a female (and vice versa at lower temperatures), TSD is favored over GSD because it enables the individual to become invariably male (or female) at those temperatures that enhance male (or female) fitness. It may appear doubtful that incubation temperature could influence an individual's characteristics years later, and no direct study of this possibility has been conducted until recently. Joanen, McNeese, and Ferguson (unpublished data) recently found that incubation temperature has significant effects on post-hatching growth rates in alligators, and the temperature which maximizes male growth differs from that which maximizes female growth. The Charnov-Bull model is thus borne out at this level, although it has yet to be shown that these differential growth rates translate into the appropriate differences in fitness.

The paradigm of ubiquitous heterogamety in amniotes has fallen, but the full implications of this new perspective remain to be demonstrated. Just how different are these mechanisms, and how is the subsequent process of sexual differentiation affected? We describe below some current work on specific aspects of these questions.

Physiology of Sex Determination. To what extent might TSD and GSD have similar physiological or biochemical bases in the control of gonad determination? This problem is being explored by application of steroid hormones to embryos to reverse sex. At present, it appears that exogenous estradiol (E) may be a general feminizer in reptiles, at least when applied at appropriate doses and appropriate times during embryogenesis. Estradiol has caused partial or complete femaleness in several species with TSD: the turtles _Emys_, _Chelydra_, and _Chrysemys_[18,19], and the leopard gecko (J.J. Bull and W.H.N. Gutzke, unpublished data). It has also been found that the feminizing effect occurs in species with GSD (or in species at least lacking TSD, when GSD cannot be demonstrated directly): the softshell turtle, _Trionyx_, the lizard, _Lacerta_, and the bullsnake, _Pituophis_ (20 for _Lacerta_; J.J. Bull and W.H.N. Gutzke, unpublished data, for the others). The administration of steroid hormones is clearly artificial, but the similar responses of all species studied may reflect similarities at some biochemical level, and E may serve as a useful probe into the natural process of sex determination.

Physiology of Sexual Differentiation. Sexuality is not a unitary phenomenon but consists of multiple phenotypes. As Money[21] pointed out, the sexuality of an individual is a cumulative composite of separate "sexes". There is the _chromosomal sex_ of the individual, i.e. the presence of heteromorphic chromosomes. The _gonadal sex_ of the individual refers to the type of gonad present and, in the majority of amniotes, the gonads of individuals with different heteromorophic chromosomes produce different gametes. The physiological sex of the individual refers principally to differences in the nature and pattern of gonadal hormone secretion between individuals of differing gonadal sex. The accessory sex structures and secondary sex characteristics together constitute the morphological sex of the individual. Probabilistic differences in behaviors between individuals of different gonadal sex are referred to as the behavioral sex of the individual. In humans, a sixth phenotype of an individual's sexuality may be considered, namely the gender identity-role, or psychological sex[22].

Clearly, in some species the multiple phenotypes of sexuality are related causally and in a similar manner[22,23]. Although relatively few vertebrates have been examined in detail, it is commonly believed that in the "typical vertebrate", fertilization sets into motion a cascade of events that culminate in the adult organism mating and contributing its genes to subsequent generations. In this scenario, the genetic constitution, and hence the chromosomal sex of the individual, is determined at conception. Genes on the sex chromosomes then channel development into either of two paths, determining the gonadal sex of the individual. Gonadal sex determination is then followed by differentiation and development of gamete-delivery systems and secondary sex characteristics, a process mediated by hormonal differences, in life, between the sexes. The gonadal hormones also influence the behaviors of each gonadal sex differentially throughout development and in adulthood.

This perspective on sexual differentiation has been developed from work on species with GSD. Does sexual differentiation under TSD work similarly? That is, do TSD and GSD differ only in that they employ different triggers of sexual differentiation? Research on leopard geckoes suggests that sexual differentiation is not the same in both systems.

Individuals predominantly exhibit reproductive behaviors characteristic of their own gonadal sex (homotypic sexual behaviors). However, individuals can, and often do, have behavior patterns typical of the opposite sex (heterotypic sexual behaviors)[24,25]. That is, while male individuals are more likely to mount and female individuals are more likely to be receptive to a mounting individual, "male" and "female" sexual behaviors simply refer to the probability that, given an appropriate stimulus, homotypic behaviors are displayed.

It is evident that the brain of diploid vertebrates is initially bisexual, and that it changes during development under the influence of the hormonal milieu. Masculinization and feminization are independent processes which, to some extent, are specific to the individual's gonadal sex. The analogy is often made that the brain mechanisms underlying sexual behaviors are like the accessory sex structures, initially bipotential and only later in development acquiring sexually dimorphic expressions. Indeed, there appear to be two distinct neural circuits that mediate sexual behavior, one which includes the anterior hypothalamus-preoptic area (AH-POA) and is involved in the control of mounting and intromission behavior, and the other which includes the ventromedial hypothalamus (VMH) and is involved in the control of sexual receptivity (Figure 1). This bimodal continuum of hormonal organization in the CNS has been demonstrated. Less is known about how this dimorphism develops either ontogenetically or historically.

In the leopard gecko (Eublepharis macularius), temperature determines sex during the middle trimester of development[10]. Incubation at high temperatures (32°C) will result in 80% of the offspring being gonadal males, whereas incubation at low temperatures (25°C) results in 100% of the individuals being gonadal females; an intermediate temperature (29°C) produces an even sex ratio (Figure 2). Recent studies of the mating behavior of these geckoes suggests that differences in mating behavior within a sex are determined partly by incubation temperature. The sexual behavior of female, but not male, leopard geckoes varied significantly, depending upon the temperature experienced during incubation[26]. Females from eggs incubated at 25°C only exhibited sexual receptivity when confronted by a male; no heterotypic behaviors were observed in tests with either male or female stimulus animals. Half of the females from eggs incubated at 29°C exhibited only homotypic behaviors to male stimulus animals, but half exhibited heterotypic behaviors in response to the courtship of stimulus males. Females from eggs incubated at 32°C frequently responded to male courtship with heterotypic behaviors. Indeed, none of these females has been observed to mate or to lay eggs, a fact which indicates that they are effectively sterile. It is significant that none of the females initiated mounting and intromission behavior, regardless of the sex of the stimulus animal, suggesting that it is the female's perception, and not her motivation, that has been influenced by incubation temperature[27]. Males, regardless of incubation temperature, did not differ in their behavior or attractivity to stimulus females.

While incubation temperature influenced the display of hetero-typical sexual behaviors only in females, agonistic behavior in both sexes was affected. Both males and females were more likely to exhibit aggressive

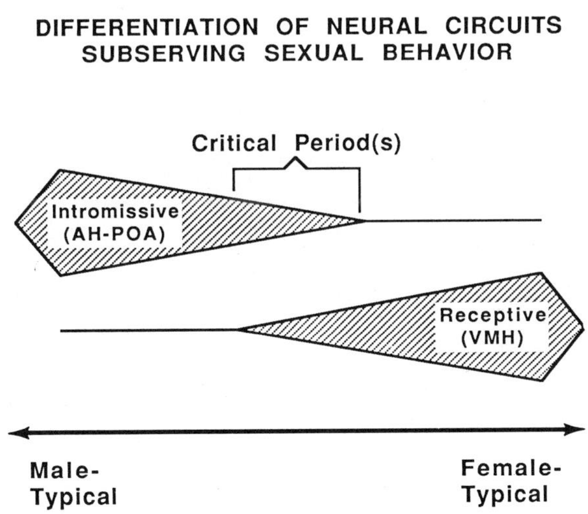

Figure 1: Schematic illustration of the current paradigm of the sexual differentiation of mechanisms in the brain which mediate sexual behavior.

behavior if they had experienced high temperatures during incubation, and both were more likely to exhibit submissive behavior if they had experienced low temperatures during incubation. Incubation temperature affected the probability of aggressive behavior by males towards other males, but not toward females. However, females were equally aggressive toward either sex.

GENDER AND THE DEVELOPMENT OF SEXUALLY DIMORPHIC BEHAVIORS

There is abundant evidence that sex steroid hormones produced by the fetal gonad, and to a lesser extent by the mother, permanently alter the brain tissues that mediate reproductive and agonistic behaviors. However, hormones during the perinatal period are not the sole determiners of psychosexual differentiation. Testicular feminization (Tfm) is a heredity defect, present in a number of mammalian species, in which genetic males develop as phenotypic females (in terms of external morphology). Although Tfm rats have an inherited insensitivity to androgen, when treated with appropriate, exogenous hormones they will exhibit mounting and intromission behavior, but no sexual receptivity[28]. This result suggests that the gender (genotype) of the individual is a major variable, independent of gonadal hormones, which must be taken into account when the organization of brain tissues that mediate sexually dimorphic behaviors is considered.

Most reptiles are gonochoristic, having separate male and female individuals; functional hermaphroditism is known only as a sporadic event within some populations. However, some species reproduce parthenogenetically[29]. In parthenogenetic species, virtually all individuals have ovaries; individuals with testes are rarely or never produced. Reproduction occurs without sperm, hence development is triggered by

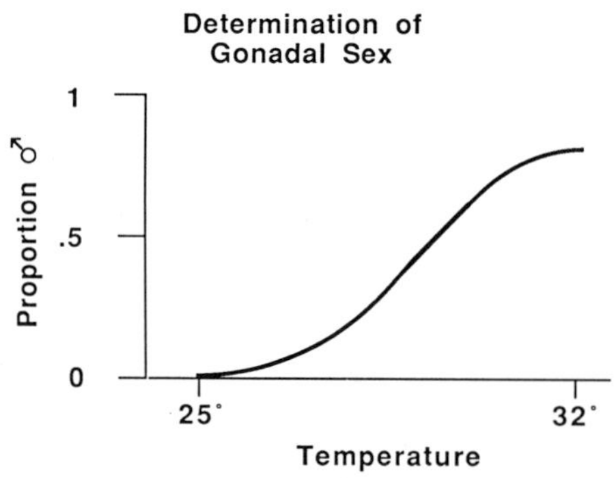

Figure 2: Temperature-dependent sex determination in the leopard gecko

some unknown mechanism. At least in one whiptail lizard, meiosis is also atypical, in that the mother's entire complement of chromosomes is duplicated prior to the first (reduction) division[30], so that each egg receives the same number of chromosomes as the mother. Each offspring is, presumably, isogenic with its mother and with its 'sibs' except for any effects of mutation and sister-chromatid exchange. As a result of this unusual genetic system, parthenogenetic species eliminate two of the greatest confounding factors in research into sex differences: individual genetic variation and gender (male genotype).

Fully one-third of the whiptail lizard species consists exclusively of females which reproduce by true parthenogenesis[29,31]. The parthenogenetic species are known to have evolved from gonochoristic species. For example, cytogenetic and electrophoretic studies indicate that the triploid parthenogen <u>Cnemidophorus uniparens</u> arose as a hybrid between recent ancestors of the two extant, diploid whiptails <u>C. inornatus</u> and <u>C. gularis</u>[32,33]. <u>Cnemidophorus inornatus</u> then backcrossed with the hybrid female, thus contributing two-thirds of the triploid genome of <u>C. uniparens</u> (L.D. Densmore, J.W. Wright, C. Moritz, W.M. Brown, unpublished data). Restriction endonuclease analyses of mitochondrial DNA indicate that the <u>C. inornatus</u> lineage is the maternal ancestor. Because representatives of both parental lineages are extant, we are presented with a natural experiment in the evolution of reproduction.

It is both remarkable and important that, although these parthenogenetic whiptails are able to reproduce without male's sperm, they still exhibit behaviors indistinguishable from the female <u>and male</u> mating behavior of their sexual ancestors[34-36]. Parthenogenetic lizards demonstrating male-like pseudosexual behavior also share a similarity in function with males of the sexually reproducing species. In sexual species, the number of ovulating females increases and the latency to ovulation decreases if a courting conspecific male is present[37]. However, a similar facilitatory effect on ovarian recrudescence also occurs in the all-female <u>C. uniparens</u> in the presence of another female who acts like a male[31,37-39]. These results indicate that behavioral facilitation of ovarian recrudescence is important in both sexual and parthenogenetic species. Elsewhere, it has been argued that this behavioral facilitation of reproduction may favor the maintenance of male-typical behaviors, perhaps accounting for the retention, in unisexual species, of the display of male-typical behavioral traits of gonochoristic species[40,41].

The whiptail lizards, therefore, provide us with an unusually appropriate animal model system with which to study the role of gender in the development of sexually dimorphic behaviors. In this regard, it is important to separate conceptually the behavioral patterns related to reproduction from the usually concomitant morphological and physiological bases of reproduction. The distinction between the psychology and the biology of sexual behavior is difficult to study, since we presently define behavioral sex by gonadal sex. Indeed, certain behaviors are regarded as secondary sex characteristics.

Some details are known about the hormonal and neural mechanisms that regulate pseudosexual behavior in parthenogenetic whiptails. Female-like pseudosexual behavior is limited to the preovulatory stage of the follicular cycle, whereas the expression of male-like pseudosexual behavior occurs most frequently during the postovulatory stages of the cycle[34,35]. Furthermore, intact individuals housed together will quickly establish and maintain a complementarity in their reproductive conditions, and alternate in their roles in pseudocopulatory encounters as they progress through their respective reproductive cycles. This alternation in behavioral role is paralleled by transitions in the circulating concentrations of sex steroid hormones produced by the ovary. Preovulatory animals, expressing female-like pseudosexual behavior, are characterized by an elevated concentration of E and a moderate concentration of progesterone (P), whereas postovulatory animals that exhibit male-like pseudosexual behavior have a two-fold lower concentration of E and a three-fold higher concentration of P[42]. Dihydrotestosterone (DHT) and testosterone (T) are not detectable in the circulation at any stage of the reproductive cycle. The surge in P following ovulation triggers male-like pseudosexual behavior in *C. uniparens*[43]. The neural circuitry involved in the control of male-like pseudosexual behavior resides in the region of the AH-POA[44]. This region is similar to the neural areas involved in the control of courtship and copulatory behavior in males of the sexual ancestral species (J.C. Rozendaal and D. Crews, unpublished data).

The parthenogenetic whiptail lizard also allows us to study other aspects of the process of sexual differentiation, such as the role of the male genotype in the hormonal induction of sexual differentiation. Since the gonads of parthenogenetic lizards always develop as ovaries, any changes in primary and secondary sex structures in the unisexual lizards are easy to identify and study. In one set of experiments the embryo's endocrine environment was manipulated to determine the consequent effects on morphology[31,45] Administration of exogenous T before and/or after hatching masculinized the gonads, accessory reproductive ducts, and secondary sex characteristics. Specifically, treatment with T affected the cross-sectional area of the gonad and the proportions of cortical and medullary tissues present in the developing gonad. The oviducts and femoral glands of T-treated individuals were hypertrophied, and the collecting tubules of the kidneys of these animals contained granules, an androgen-dependent, sexually dimorphic characteristic present in male, gonochoristic squamate reptiles.

ROLE OF SEX HORMONES IN THE ORGANIZATION OF NEURAL CIRCUITS THAT MEDIATE TO SEXUAL BEHAVIOR

There is enormous diversity among the vertebrates in the nature and pattern of reproduction[46,47] This diversity reflects an equally great diversity in the mechanisms that control reproduction. To understand the relationship between the organism's environment, its pattern of reproduction, and the mechanisms subserving reproduction, it is useful to recognize that three categories of constraints have shaped reproductive processes and, hence, the underlying mechanisms: environmental, physiological/developmental, and phylogenetic[38]. In this context the reproductive process can be divided into three major components: the production of gametes, the secretion of sex steroid hormones by the gonad, and the timing of mating behavior. Recent studies indicate that the relationship between these three components can differ profoundly, depending upon the constraints that are operating on the individual. For example, most homeotherms appear to exhibit an *associated* reproductive pattern in which there is a temporal association between the production of gametes, increasing concentrations of sex steroid hormones, and the exhibition of mating behavior[47]. Ectotherms, particularly those living in harsh environments, exhibit a dissociated reproductive pattern, because it is often metabolically impossible for the gametes to develop prior to the brief breeding season. In such animals, gonadal recrudescence and the corresponding rise in levels of sex hormones occurs only after mating has stopped; the gametes produced are then stored until the next reproductive period. How then do species exhibiting this dissociated reproductive pattern mate at a time when gonadal sex steroid hormones are at their lowest concentration?

The prevailing paradigm for research in the sexual differentiation of brain and behavior is that sex steroid hormones exert an organizing effect early in life on the brain tissue that mediates sexual behaviors[48]. Goy and McEwen[23] proposed three types of relationship between hormones and sexually dimorphic behavioral traits. Type I traits encompass those behaviors that require the relevant hormones both early in development and later in life for their expression. Type II behavioral traits include those behaviors that require the relevant hormones for activation of the behavior only at a later age but are, apparently, independent of organizational influences of hormones. Type III behaviors are the complement of Type II behaviors, in that they require hormones during an early critical period but are independent of hormonal control after puberty. Examples of Type III behavior are few: rough-and-tumble play and mounting behavior in rhesus monkeys[49]; mounting[50] and aggressive behavior[51] in female guinea pigs; and the mictuitional patterns of dogs[52,53].

Our present understanding of the organizational effects of sex steroid hormones on neural circuits in reptiles is rudimentary, and is derived from studies on the red-sided garter snake, *Thamnophis sirtalis parietalis*. In the red-sided garter snake, mating behavior in the adult is independent of control by gonadal sex steroid hormones[54,55]. Courtship behavior will occur only if animals are housed under warm conditions after a prolonged cold period of dormancy. In nature, courtship behavior in the male is activated by the springtime increase in temperature[56], an event easily simulated in the laboratory[55,57].

While gonadal sex hormones do not appear to activate mating behavior in the adult red-sided garter snake, they do exert an effect before adulthood. Castrated neonatal snakes, treated with T, courted on emergence from their first hibernation, whereas castrated neonates which had not received androgen treatment as neonates ignored females[58]. This result suggests that, in the red-sided garter snake, sex steroid hormones act to organize the neural circuits which subserve courtship behavior, such that fluctuations in temperature, and not increasing concentrations of sex steroid hormones, activate sexual behavior in the adult organism. This type of effect would be catalogued as a Type III behavioral trait.

ALTERNATIVE LIFE HISTORIES WITHIN A SPECIES: UNEXPLORED PROBLEMS IN SEXUAL DIFFERENTIATION

It is now well-established that, in many species, individuals of the same gender within a given population may exhibit a mixture of sexual phenotypes, and may display heterotypical physiologies, morphologies, and/or behaviors[59]. Furthermore, individuals exhibiting these alternative reproductive strategies are successful reproducers. Two examples are given below.

Cryptic Males in Fish. There are three phenotypes of male bluegill sunfish (*Lepomis macrochirus*): a large, colorful male that defends territories and solicits females; a small male that sneaks matings when the territorial male is otherwise occupied; and a large but drab male that mimics females in appearance[60]. The mimics effectively insert themselves between a courting territorial male and the female he is courting and, in this manner, the mimic fertilizes the female. These three phenotypes represent two developmental pathways: the smallest type of male develops into a mimic of females, but the mimic does not go on to become a large, territorial male (this latter pathway is separate). The maintenance of multiple male strategies poses a number of interesting questions for theories of sexual differentiation. What are the causal mechanisms and how do the mechanisms which produce different strategies compare with each other?

'She-Males' in Snakes. Recent studies of the red-sided garter snake have also revealed that some males mimic females. Courtship in adult, male red-sided garter snakes is elicited by an attractiveness pheromone on the skin of females, and a receptive female in an area with a high density of males often generates a large congregation of courting males, known as a mating ball[62-64]. In one Canadian locality, in 16% of more than 200 'mating balls' in which the individuals were censused, a male was the object of courtship of the mating aggregations, and no female was observed[61]. Since mating balls typically possess 20 or more males, the actual frequency of these mimics of female snakes in the population is probably less than 1%. These males (termed she-males) release a pheromone that has similar, if not identical, properties to the attractiveness pheromone released by adult females.

These mimics of females appear to have a decided mating advantage. In competitive mating trials, in which normal males and a single she-male were placed in an arena with an unmated female, the she-male mated with the female significantly more often than did the normal males[61]. It appears that the she-male confuses the other males by providing a second source of pheromone in the mating ball.

Behavioral, morphological, and physiological studies reveal that she-males are genetic males which do not differ significantly from other males in length and weight[65]. Furthermore, the she-males have fully functional testes and accessory sex structures identical to those of other males. They will court and mate with females, but it is not known as yet whether they can father offspring. She-males appear to be physiologically feminized. Radioimmunoassay of the concentration of circulating sex steroid hormones indicates that she-males have very high levels of T (Table 1). Although circulating levels of E are similar to those found in normal males, it is possible that aromatase enzymes which convert T into E in the liver and skin, the target organs involved in the production and release of the attractiveness pheromone[64], produce high local concentrations of E which effectively feminize the male.

The she-male trait could have an environmental and/or genetic basis. One possible mechanism for an environmental (non-inherited) basis involves hormonal induction. A precedent established from work on rodents is that fetal position in the uterus has an influence on both the morphology and behavior of the adult; males which develop between two females are feminized as compared to males which develop between two males[66].

Table 1: Circulating levels of sex steroid hormones in the two male morphs and the female red-sided garter snake. Values are the mean ng/ml + s.e.m. (From Mason and Crews, 1985).

	(n)	Hormone Testosterone	Estradiol	Dihydro-testosterone
Male	9	5.72 ± 1.65	0.06 ± 0.01	0.23 ± 0.07
Female	9	0.23 ± 0.08	7.70 ± 3.44	0.09 ± 0.02
She-male	9	19.78 ± 5.70	0.08 ± 0.01	0.14 ± 0.02

A similar phenomenon might occur in the viviparous, red-sided garter snake, although no studies have specifically investigated this possibility. Since she-males both produce and release the attractiveness pheromone, this feminization must be physiological as well as morphological. Such feminization could be brought about by estrogens early in life (perhaps from female neighbors during development). This hypothesis is supported by the observation that administration of estrogen to neonatal males results in their eliciting courtship from adult males[58]. Similar treatment of adult males does not increase their attractiveness to other males (indicating that their skin cannot release the pheromone)[63]. Exogenous E does, however, stimulate production of the attractiveness pheromone[63]. She-males are similar to females in that the pheromone is actively transported from the blood to the surface of the skin.

CONCLUSIONS

In vertebrates, observed similarities in the deterministic relationships among the different phenotypes of sexuality have been assumed to reflect a phylogenetically stable core system. However, this apparent conservation of physiology may reflect more the nature of the particular organisms studied to date rather than fundamental characteristics of vertebrates. Just as recent studies are questioning the generality of certain hormone-behavior relationships of vertebrates, the diversity in the expression of sexuality in vertebrates is clear proof that intrinsic functional associations among the various sexual phenotypes can differ considerably.

Many vertebrates simply do not fit the "typical vertebrate" model of sexuality. Instead of genetic sex determination, there are species in which sex is determined by the embryonic environment. Rather than separate sexes in separate individuals (gonochorism), there are species in which both types of gonad reside in a single individual (hermaphroditism) either sequentially or simultaneously. In still other species, reproduction is by parthenogenesis, rather than by gametic union. Even in gonochoristic species, the sexes may not appear to differ in the nature or in the pattern of gonadal hormones secreted, nor do they possess specialized duct systems for transporting the gametes to the external environment. Furthermore many species are sexually monomorphic externally, whereas in other species some individuals of one sex may mimic the opposite sex. Finally, it is now well established that, during mating, each sex exhibits both heterotypic and homotypic sexual behaviors.

These experiments of nature need to be utilized to uncover the true generalities in sex determination and sexual differentiation. This article described how, in addition to providing certain natural experiments with which to probe the causal mechanisms of sex determination and sexual differentiation, reptiles also afford opportunities for the combination of studies of mechanisms with studies of the functional outcomes of these processes.

ACKNOWLEDGEMENTS

We thank Cecilia Martinez-Veliz and Kristy Vick for typing the manuscript. Original research described in this chapter was supported by NICHHD 16687, The Whitehall Foundation, and NIMH Research Scientist Development Award MH00135 to David Crews and by NSF MSR 84-15745 to James J. Bull.

REFERENCES

1. Licht, P: Suitability of the mammalian model in comparative reproductive endocrinology. In: Ralph, CL, ed. Comparative Endocrinology. New York: AR Liss, Inc. 95-114, 1986.

2. Witschi, E: Age of sex determining mechanisms in vertebrates. Science, 130:372-375, 1959.

3. Ohno, S: Sex Chromosomes and Sex-linked Genes. Berlin: Springer, 1967.

4. Charnier, M: Action de la temperature sur la sex-ratio chez l'embryon d'Agama agama (Agamidae, Lacatilien). Societie Biologie Ouest Africa 160:62-622, 1966.

5. Pieau, C: Sur la proportion sexuelle chez les embryons de deux Cheloniens, (Testudo graca L. et. Emys orbicularis L.) issus d'oeufs incubes artificiellement. CR Acad Sci, 272:3071-3074, 1971.

6. Bull, JJ: Sex determination in reptiles. Quart Rev Biol, 55:3-21, 1980.

7. Ferguson, MWJ, Joanen, T: Temperature of egg incubation determines sex in Alligator mississippiensis. Nature, 292:850-853, 1982.

8. Bull, JJ: Evolution of Sex Determining Mechanisms. Menlo Park: Benjamin/Cunmmings. 1983.

9. Bull, JJ: Temperature dependent sex determination in reptiles: Validity of sex diagnosis in hatchlings. Can J Zool, (in press).

10. Bull, JJ: Temperature sensitive periods of sex determination in a lizard: comparisons with turtles and crocodilians. J Exp Zool, (in press).

11. Mrosovsky, N, Hopkins-Murphy, SR, Richardson, JI: Sex ratio of sea turtles: seasonal changes. Science, 225:739-741, 1984.

12. Wilhoft, DC, Hotaling, E, Franks, P: Effects of temperature on sex determination in embryos of the snapping turtle, Chelydra serpentina. J. Herpetol, 17:38-42, 1983.

13. Vogt, RC, Bull, JJ: Ecology of hatchling sex ratio in map turtles. Ecology, 65:582-587, 1984.

14. Webb, GJW, Buckworth, R, Manolis, SC: Crocodylus johnstoni in the McKinlay River, N.Y. VI. Nesting Biology. Australia J Wildlife Res, 10:607-637, 1983.

15. Bull, JJ: Sex ratio and nest temperature in turtles: Comparing field and laboratory data. Ecology, 66:1115-1122, 1985.

16. Schwarzkopf, L, Brooks, RJ: Sex determination in northern painted turtles; effect of incubation at constant and fluctuating temperatures. Can J Zool, 63:2543-2547, 1985.

17. Charnov, EL, Bull, JJ: When is sex environmentally determined? Nature, 266:828-830, 1977.

18. Pieau, C: Differenciation du sexe en fonction de la temperature chez les embryons d'Emys orbicularis L. (Chelonien). Effets des hormones sexuelles. Ann Embryol Morph, 7:365-394, 1974.

19. Gutzke, WHN, Bull, JJ: Steroid hormones reverse sex in turtles. Gen Comp Endocr, 1986 (in press).

20. Raynaud, A, Pieau, C: Embryonic development of the genital system. In: Gana, C, Billett, F, eds. Biology of Reptilia, Vol 15 Development B. New York: John Wiley and Sons, 149-300, 1985.

21. Money, J: Hermaphroditism, gender, and precocity in hyperandrenocorticism: Psychological findings. Bull Johns Hopkins Hosp, 96:253-264, 1955.

22. Money, J: Intersexual and transexual behavior and syndromes. In: Arieti, S, Brady, EB, eds. American Handbook of Psychiatry, Volume 3. New York: Basic Books, 253-264, 1974.

23. Goy, RW, McEwen BS: Sexual Differentiation of the Brain. MIT Press, Cambridge: MIT Press, 1980.

24. Beach, FA: Factors involved in the control of mounting behavior by female mammals. In: M Diamond, ed. Perspectives in Reproduction and Sexual Behavior, Bloomington: Indiana University Press, 83-132, 1968.

25. Dagg, AI: Homosexual behavior and female-male mounting in mammals. Mamm Rev, 14:155-185, 1984.

26. Gutzke, WHN, Crews, D: Incubation temperature affects sexual and agonistic behavior of adult leopard geckos. Amer Zool 1986 (in press).

27. Beach, FA: Hormones and psychological processes. Can J Psychol 37:193-210, 1983.

28. Shapiro, BH, Levine DC, Adler NT: The testicular feminized rat: A naturally occuring model of androgen-independent brain masculinization. Science 209:418-420, 1980.

29. Cole, CJ. Evolution of parthenogenetic species of reptiles. In: Reinboth, R. ed: Intersexuality in the Animal Kingdom, New York: Springer-Verlag 340-355, 1975.

30. Cueller, O: Reproduction and the mechanism of meiotic restitution in the parthenogenetic lizard, Cnemidophorus uniparens. J Morph 133:139-166, 1971.

31. Crews, D, Gustufson, JE, Tokarz, RR: Psychology of parthenogenesis in reptiles. In: Huey, R, Pianka, E, Schoener, TW, eds. Lizard Ecology. Cambridge: Harvard University Press, 205-231, 1983.

32. Lowe, J, Wright, TW. Evolution of parthenogenetic species of Cnemidorphorus (Whiptail) lizards in Western North America. J Ariz Acad Sci, 4:81-87, 1966.

33. Neaves, WB: Adenosine deaminase phenotypes among sexual and parthenogenetic lizards in the genus Cnemidophorus. J Exp Zool, 171:174-183, 1966.

34. Crews, D, Fitzgerald, K: "Sexual" behavior in parthenogenetic lizards (Cnemidophorus). Proc Natl Acad Sci, 77:499-502, 1980.

35. Moore, MC, Whittier, JM, Billy, AJ, Crews, D: Male-like behavior in an all-female lizard: Relationship to ovarian cycle. Anim Behav, 33:284-289, 1985.

36. Lindzey, J, Crews, D: Hormonal control of courtship and copulatory behavior in male Cnemidophorus inornatus, a direct sexual ancestor to the parthenogen, C. Uniparens. Gen Comp Endocr, 64:411-418, 1986.

37. Crews, D, Grassman, M, Lindzey, J: Behavioral facilitation of reproduction in sexual and parthenogenetic whiptail (Cnemidophorus) lizard. Proc Nat Acad Sci, 1986 (in press).

38. Crews, D, Moore, MC: Evolution of mechanisms controlling mating behavior. Science 231:121-125, 1986.

39. Gustafson, JE, Crews, D: Effect of group size and physiological state of a cagemate on reproductive effort in the parthenogenetic lizard, Cnemidophorus uniparens (Teiidae). Behav Ecol Sociobiol, 8:267-272, 1981.

40. Crews, D, Teramoto, J, Carson, HL: Behavioral facilitation of reproduction in sexual and parthenogenetic Drosophila. Science, 227:77-78, 1985.

41. Crews, D: Diversity and evolution of behavioral controlling mechanisms. In Crews, D, ed. The Psychobiology of Reproductive Behavior: An Evolutionary Perspective. New Jersey: Prentice-Hall, Inc., 88-119, 1987.

42. Moore, MC, Whittier, JM, Crews, D: Sex steroid hormones during the ovarian cycle of an all-female, parthenogenetic lizard. Gen Comp Endocr, 60:144-153, 1985.

43. Grassman, M, Crews, D: Progesterone induction of male-like pseudosexual behaviors and stimulus-response complementarity in an all-female lizard species. Horm. Behav, 20:327-335, 1986.

44. Mayo, M, Crews, D: Neural control of male-like pseudocopulatory behavior in the all-female lizard, Cnemidophorus uniparens: Effects of intracranial implantation of dihydrotestosterone. Horm Behav. (in press).

45. Billy, A, Crews, D: The effects of sex steroid treatments on sexual differentiation in a unisexual lizard, Cnemidophorus uniparens (Teiidae). J Morph, 18:129-142, 1986.

46. Blackwelder, RE, Shepard, BA: The Diversity of Animal Reproduction. Boca Raton: CRC Press, 1981.

47. Crews, D: Gamete production, sex hormone secretion, and mating behavior uncoupled. Horm Behav, 18:22-28, 1984.

48. Phoenix, CH, Goy, RW, Gerall, AA, Young, WC: Organizational action of prenatally administered testosterone propionate on the tissues mediating behavior in the female guinea pig. Endocr, 65:369-382, 1959.

49. Goy, RW: Early hormonal influences on the development of sexual and sex-related behavior. In: Schmidt, FO, ed. The Neurosciences. Second Study Program. Cambridge: MIT Press, 196-206, 1970.

50. Goy, RW, Bridson, F, Young, WC: Period of maximal susceptibility of the prenatal guinea pig to masculinizing action of testosterone propionate. J Comp Physiol Psychol, 57:166-174, 1964.

51. Goldfoot, DA: Sex-specific behavior-specific actions of dihydrotestosterone: Activation of aggression, but not mounting in ovariectomized guinea pigs. Horm Behav, 13:241-255, 1979.

52. Martins, T, Valle, SR: Hormonal regulation of the micturitional behavior of the dog. J Comp Physiol Psychol, 41:301-311, 1948.

53. Beach, FA: Effects of gonadal hormones on urinary behavior in dogs. Physiol Behav, 12:1005-1013, 1974.

54. Crews, D, Garstaka, WR: The ecological physiology of reproduction of the garter snake. Sci Amer, 247:158-168, 1982.

55. Crews, D: Control of male sexual behavior in the Canadian red-sided garter snake. (Thamnophis sirtalis sirtalis). Horm Behav, 7:451-460, 1983.

56. Aleksiuk, M, Gregory, PT: Regulation of seasonal mating behavior in Thamnophis sirtalis parietalis. Copeia, 1974:681-689, 1974.

57. Camazine, B, Garstka, W, Tokarz, R, Crews, D: Effects of castration and androgen replacement on male courtship behavior in the red-sided garter snake (Thamnophis sirtalis parietalis). Horm Behav, 14:358-372, 1980.

58. Crews, D: Effects of early sex steroid hormone treatment on courtship behavior and sexual attractivity in the red-sided garter snake (Thamnophis sirtalis parietalis). Physiol Behav, 35:569-575, 1985.

59. Austad, SN, Howard, RD: Alternative reproductive tactics. Amer Zool, 24:305-418, 1984.

60. Gross, MR: Sunfish, salmon, and the evolution of alternative reproductive strategies and tactics in fishes. In: Potts, GW, Wooten, RJ, eds, Fish Reproduction: Strategies and Tactics. New York: Academic, 1984: 55-75.

61. Mason, RT, Crews, D: Female mimicry in garter snakes. Nature, 316:59-60, 1985.

62. Crews, D: Hormonal control of male and female sexual behavior the garter snake (Thamnophis sirtalis sirtalis). Horm Behav, 7:451-460, 1976.

63. Garstka, WR, Crews, D: Female sexual pheromone in the skin and circulation of a garter snake. Science, 214:681-683, 1981.

64. Mason, RT, Chinn, JW, Crews, D: Skin lipids of garter snakes serve as semiochemicals. Ann NY Acad Sci, 1987 (in press).

65. Mason, RT, Crews, D: Pheromonal mimcry in garter snakes. In: Duval, D, Muller-Schwartz, D, eds. Chemical Signals and Vertebrates. New York: Plenum Press, 1986 (in press).

66. VomSaal, FS: The interaction of circulating oestrogens and androgens in regulating mammalian sexual differentiation. In: Balthazard, J, Prove, E, Gilles, R, eds. Hormones and Behavior in Higher Vertebrates, Berlin: Springer-Verlag, 1983.

SEXUAL DIFFERENTIATION IN MARSUPIALS

M.B. Renfree, G. Shaw and R.V. Short

Departments of Anatomy and Physiology
Monash University
Melbourne, 3168
Australia

INTRODUCTION

It is most appropriate that a Workshop on Sexual Differentiation should include a review of the situation in marsupials, since they are distinguished from eutherian mammals not only by their different mode of reproduction, but also by the nature, timing and extent of the sexual differentiation of their external and internal genitalia.

As the name implies, most female marsupials possess a pouch or marsupium on the ventral abdominal wall that encloses the teats and mammary glands, although in several genera, such as *Monodelphis*, the pouch is absent. The pouch and teats are absent in all adult males. There has been much debate as to whether the pouch and the scrotum are homologous structures, akin to the labia majora and scrotum of eutherian mammals, and it is unclear whether or not their development is controlled hormonally[25]. Unlike the analogous structures in most eutherians, except possibly the beaver, the rectal, urinary and genital passages of all marsupials open into a common urogenital sinus with a single external orifice, through which the penis of the male protrudes when erect.

Internally, all female marsupials possess two separate uteri, which open by separate cervices into the vaginal culs-de-sac[25]. Two vaginae run lateral to the ureters to open posteriorly into the urogenital sinus; these lateral vaginae are used exclusively for sperm ascent. Birth occurs through a third structure, the median vagina or birth canal, which also connects the vaginal culs-de-sac with the urogenital sinus. This median vagina becomes patent at the time of parturition and closes up again afterwards in most species, although it remains patent after the first parturition in the macropodids (kangaroos and wallabies). Male marsupials are unlike eutherians in that their scrota are anterior to the penis, and their vasa deferentia open into the urethra without first having to curve around the ureters[25].

In all the species of marsupials that have been investigated, a mere handful of the 249 extant species, it is difficult to sex the offspring at birth by external phenotypic characteristics[25]. Gonadal differentiation mainly takes place after birth, something that is unheard of in eutherians, and is followed by differentiation of the Wolffian and Mullerian ducts, and testicular migration and descent.

Pioneers in the study of mammalian sexual differentiation, such as R.K. Burns and C.R. Moore, realized that marsupials provide a unique opportunity for studying sexual differentiation in what amounts to an exteriorized fetus. The marsupial neonate is not under the influence of any placental hormones at the time of gonadal differentiation, and this independence suggests that it might be possible to modify the gonads by administration of sex steroids, as is the case in fish and amphibians, where steroids can induce a complete, functional sex reversal. Steroids have no effect on gonadal differentiation in eutherian mammals, a characteristic which evolved perhaps as a defense to protect the fetus from the influence of placental steroids.

Another advantage of marsupials for studies of sexual differentiation is that they uniformly have a small number of large chromosomes, making it relatively easy to distinguish the sex chromosomes[18]. It is a simple matter to karyotype pouch young from a biopsy of the tip of the tail, and it is clearly essential to establish the genetic sex when studying early gonadal differentiation.

Although early investigators in the United States realized the experimental advantages of marsupial pouch young, which are readily accessible for administration of hormones or surgical manipulation during the critical stages of sexual differentiation, the only marsupial available to them locally, indeed the only marsupial whose range extends into North America, is the omnivorous Virginia opossum, *Didelphis virginiana*. Although it has the advantage of being polytocous, with a litter size of 3-13, it is a seasonal breeder, and even by marsupial standards it has a very short gestation (13 days). The newborn young weigh a mere 125 mg, and are too small for experimental convenience. After the Second World War, work on *Didelphis* virtually ceased, and our understanding of its basic endocrinology now lags far behind our understanding of that of some of the Australian marsupials.

The best understood of Australian marsupials is the tammar wallaby, *Macropus eugenii*[25], a medium-sized (5 kg) herbivorous macropodid now largely confined to islands off the south and south-west coasts of Australia. It is also a seasonal breeder, as befits its southerly latitude, normally giving birth to a single offspring, which weighs about 450 mg, approximately a month after the summer solstice, with pouch exit late in the following spring, about 9-10 months later.

Tammars are easy to keep and breed well in captivity. They have the great experimental advantage that births can be induced at almost any time of the year, and timed in advance to take place almost to the day. This useful characteristic is due to the fact that the tammar undergoes embryonic diapause[25]. Within a day of giving birth, the female comes into oestrus, mates, and ovulates, but the sucking activity of the newborn young in the pouch inhibits the development and secretory activity of the newly formed corpus luteum, thereby holding uterine secretions in check and arresting the development of the embryo at the blastocyst stage, when there are about 100 cells present. This phase of lactational diapause normally lasts for 4 or 5 months. During this time, luteal hypertrophy can be induced and the blastocyst reactivated simply by removal of the pouch young from the teat, resulting in a new birth 26-27 days later[25].

During the winter, lactational diapause gradually gives way to seasonal diapause, when the development of the corpus luteum is held in check by the prevailing photoperiod, an effect that is mediated via pineal secretion of melatonin. Seasonal diapause can usually be terminated by denervating or removing the pineal gland, or by giving a subcutaneous implant or timed daily injections of melatonin, which will result in the birth of young 27-35 days later[25].

These simple experimental procedures of removal of pouch young or pineal manipulation have added enormously to our ability to study marsupial sexual differentiation, since we can produce pouch young of known age to order. Furthermore, the size of the tammar neonate makes it most suitable for experimental surgery.

The purpose of this paper is to review what is known about sexual differentiation in the Virginia opposum, and to compare it with new information we and others have obtained from studies of the tammar wallaby.

SEXUAL DIFFERENTIATION OF THE POUCH YOUNG

The Virginia Opossum

Figure 1 summarizes what is known about the sequence of events during sexual differentiation in the Virginia opossum, and is based largely on the work of Hartman, Morgan, Moore, McCrady and Burns.

The gonads: At birth, the gonads are undifferentiated in both sexes, with a homogeneous medullary blastema surrounded by a single layer of coelomic epithelial cells[20,22]. By day 1 of pouch life, the primary sex cords appear, arranged centripetally within the blastema, and they contain occasional, large, clear cells, presumably the primordial germ cells[13,20,22].

The first sign of testicular differentiation is seen on day 3-4, when the primary sex cords become well developed, and surrounded by a primary tunica albuginea covered by a single layer of coelomic epithelium[10]. Primordial germ cells are confined to the central regions of the developing testis.

Ovarian development lags several days behind that of the testis[22]. The primary sex cords are less well defined, and the epithelial layer is 2-3 cells thick. The majority of primordial germ cells migrate into the gonads during the first week of pouch life, and mitotic activity in the female germ cells is maximal on about day 24[22].

The testes begin their trans-abominal migration soon after day 14 and enter the inguinal canal at about day 30. Testicular descent into the scrotum is completed by about day 80[16].

Recent studies *in vitro*[17] have shown that the gonads become steroidogenically active early in their differentiation. Pooled gonads from young of unknown sex, and less than 10 days old, produced progesterone and testosterone from radiolabelled precursors. By day 19-26 onwards there was evidence of a sex difference in steroidogenic capacity; the testes but not the ovaries produced testosterone, whereas the ovaries but not the testes had significant aromatase activity.

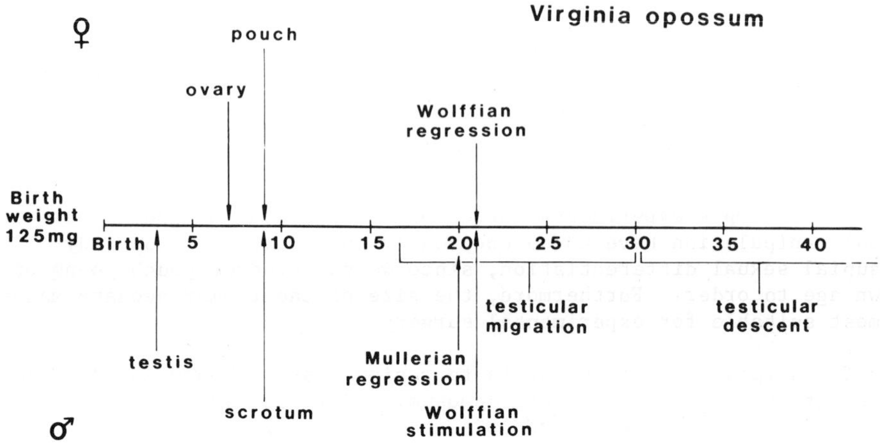

Figure 1: Summary of the sequence of events during sexual differentiation of the pouch young of the Virginia opossum, Didelphis virginiana. Data from various authors, see text.

The Wolffian ducts: At birth, there is a functional mesonephric kidney, but by day 10 it has begun to degenerate as the metanephric kidney takes over, and by day 16-17 mesonephric degeneration is advanced[19,22]. As mesonephric regression takes place, some of the most anterior mesonephric tubules appear to invade the hilus of the gonad, to form the rete tubules. These tubules form a clearly distinguishable band of dark-staining tissues in both the testis and ovary[8,9,22]. In the male, they become the ductuli efferentes, connecting the testis with the epididymis, which differentiate from between 5 and 8 of the more anterior mesonephric tubules. In the female, the rete begins to degenerate after day 20, and is vestigial by day 50, although remnants may be seen in the adult ovary[22]. The main mesonephric or Wolffian duct develops identically in males and females up to day 20, but thereafter, in the male, it enlarges and differentiates into the vas deferens, and by day 50 there is a well-developed epididymis with caput, corpus and cauda[7]. In females, the Wolffian duct regresses after day 20, and by day 50 it persists only as a discontinuous solid cord of cells[7,12].

George et al.[17] have shown high 5α-reductase activity in the mesonephric ducts of pouch young of less than 10 days old, but the activity becomes virtually undetectable from days 19 to 75 when development of the Wolffian duct is taking place, a result which suggests that development of this structure is more likely to be induced by testosterone itself, rather than by 5α-dihydrotestosterone.

The Mullerian ducts: The Mullerian or paramesonephric ducts first appear as an invagination of the coelomic epithelium on the cranial pole of the mesonephros about 3-4 days post-partum[10,11,20]. The duct tunnels into the border of the mesonephros to form a solid cord of cells parallel and lateral to the Wolffian duct. It crosses the posterior end of the Wolffian duct ventrally, reaching the urogenital sinus by day 10. In females, the posterior end has become patent by about day 17.

By day 20, the Mullerian duct is completely formed in both sexes[7]. From then onwards, the duct begins to involute in males, with progressive occlusion which commences at the caudal end, although there is histological evidence of simultaneous degeneration of the ostium abdominale at the anterior end[10,11]. By day 50, only a few anterior segments of the duct still retain a lumen. The vaginal complex starts to form in females at about day 22, and by day 50 it is complete, with lateral and median vaginae, two uteri, and coiled oviducts that end in fimbriae which invest the ovaries.

Nipples, pouch and scrotum: There are usually about six pairs of mammary anlagen and nipples present in both sexes at the time of birth (C.G. Hartman, cited in ref. 7). They form a crescent which extends from just anterior to the umbilicus, uniting posteriorly just in front of the genital tubercle. There is often an additional single, unpaired median nipple near the posterior end of the series. The nipples are surrounded by paired, bilateral pouch folds in both sexes.

By 10-12 days after birth, sex has become distinguishable externally by the appearance of scrotal anlagen just cranial to the genital tubercle[10,19]. McCrady[19] claims that the scrotum is formed by fusion of the posterior ends of the pouch folds. The pouch folds and nipples then regress in the male.

Urogenital sinus and external genitalia: Prostatic buds develop from the urogenital sinus and are only found in the male. They start to appear on day 16-18[12,21], and by day 50 the prostate is fully formed. In both sexes there is a prominent phallus at birth. It enlarges progressively in males after day 15, to form the characteristic bifid glans penis.

The tammar wallaby

In keeping with its greater relative body weight at birth, the tammar neonate is more advanced in its development than the Virginia opossum. The principal events of sexual differentiation in the tammar pouch young are summarized in Figure 2 which is based on the work of Alcorn[1,2] and our own unpublished results.

The gonads: On the basis of a small number of observations, Alcorn is of the opinion that the gonads may already show some indication of sexual dimorphism at the time of birth, when the testis may be slightly larger than the ovary and have more germ cells within the sex cords of the central somatic blastema. However, more evidence is required to establish this point. Dividing germ cells are seen in the testis soon after birth, and testicular development in the first few days of life is rapid, with fewer but larger sex cords appearing by day 3, and clearly developed seminiferous tubules with patent lumina present by day 7. Between days 7 to 14, the testis becomes quite round in appearance, while the ovary remains much more elongated and fusiform (see Figures 3 and 4).

In the female at birth, the ovary has not started to differentiate, but by 12 hours after birth, germ cells appear to be migrating from the centre of the gonad to its periphery[1,2]. By day 3, the number of germ cells has increased, and more of them are concentrated at the periphery of the gonad; by day 5 the central medulla near the hilus is

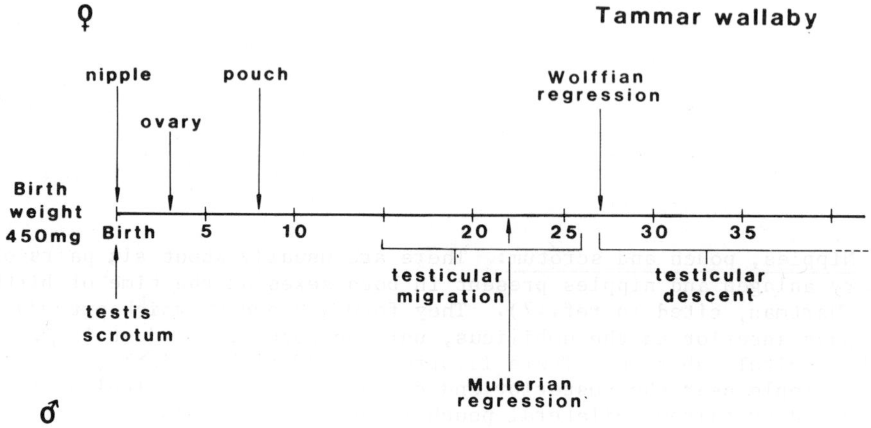

Figure 2: Summary of the sequence of events during sexual differentiation of the pouch young of the tammar wallaby, *Macropus eugenii*. Data from Alcorn[1,2] and our own observations.

virtually devoid of germ cells. By day 8, germ cells are concentrated in the outer cortex which is still incompletely separated from the medulla by a thin, discontinuous stromal layer. The medulla consists of poorly organized cords from which Alcorn believes the *rete ovarii* is formed.

Between days 12 and 22, the number of female germ cells increases markedly as they undergo 2 or 3 mitotic divisions, so that by day 22 the cortex contains numerous nests of primordial germ cells[2]. These nests of germ cells enter meiosis synchronously, starting on about day 24 and continuing up to day 190[2]. The first evidence of formation of primordial follicles is seen on about day 110[2]. The origin of the follicular cells is unclear. In another marsupial, the bandicoot, the rete plays no part in the formation of follicles, whereas in the potoroo, bettong and brush possum, it appears to contribute to folliculogenesis[26].

The testes start to pass through the inguinal canal on about day 30, and enter the neck of the scrotum on about day 40. Scrotal descent is complete around day 55-60.

The Wolffian ducts: At birth, there is a functional mesonephric kidney, so the Wolffian ducts are patent and functional urinary ducts in both sexes. In the male, they differentiate into the epididymis and *vas deferens*, whereas in the female they become degenerate by day 31[1].

The Mullerian ducts: In the neonatal female, a short patent Mullerian duct is present and the *ostium abdominale* opens just in front of the mesonephros on the anterior, free margin of the urogenital fold. In the neonatal male, the Mullerian duct is longer than in the female. By day 3 in the female, the duct is patent for the full length of the urogenital fold, but ends blindly in the urogenital sinus, whereas in the male, the duct is only half-developed[1]. By day 8, the Mullerian duct of the male has opened into the Wolffian duct at its point of juncture with the urogenital sinus, but this event does not occur until

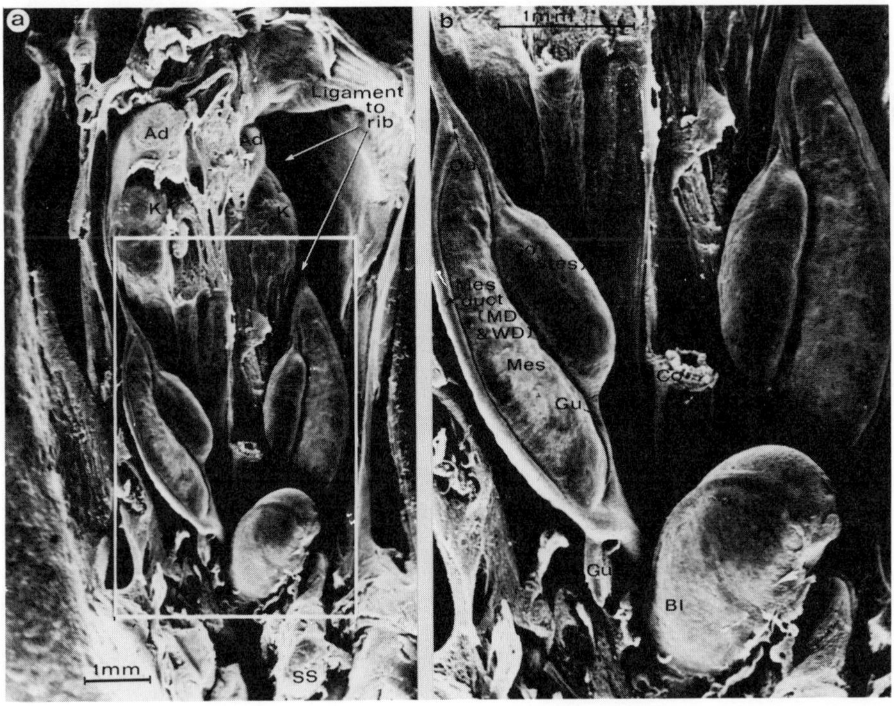

Figure 3: Scanning electron micrograph of the internal genitalia of an 8-day-old male tammar pouch young. (a) Low power magnification of whole abdominal cavity. (b) High power magnification of inset shown in (a). Ad, adrenal gland; Bl, bladder; Co, colon; Go, gonad; Gu, gubernaculum; K, metanephric kidney; MD, Mullerian duct; Mes, mesonephros; Mes duct, Wolffian and Mullerian (paramesonephric) ducts; Oa, ostium abdominale; SS, scrotal swelling; WD, Wolfian duct.

day 17 in the female[1]. By day 22, the Mullerian duct of the male has started to regress.

<u>Nipples, pouch and scrotum:</u> Four nipple anlagen are present at the time of birth in females, as histologically detectable keratinous thickenings of the abdominal skin, but they are apparently absent in males[1].

The pouch appears to be represented by short, paired, bilateral folds that fuse posteriorly to cover the nipple anlagen, and it first becomes visible in the female at about 8 or 9 days of age.

The scrotum arises from short, paired, bilateral swellings just anterior to the phallus, and Alcorn[1] has described the appearance of scrotal anlagen in an embryo 25 days after removal of the pouch young, one to two days before expected birth; we have observed scrotal anlagen in an embryo 23 1/2 days after removal of pouch young. Alcorn concludes that males can be correctly sexed at birth by the presence of scrotal

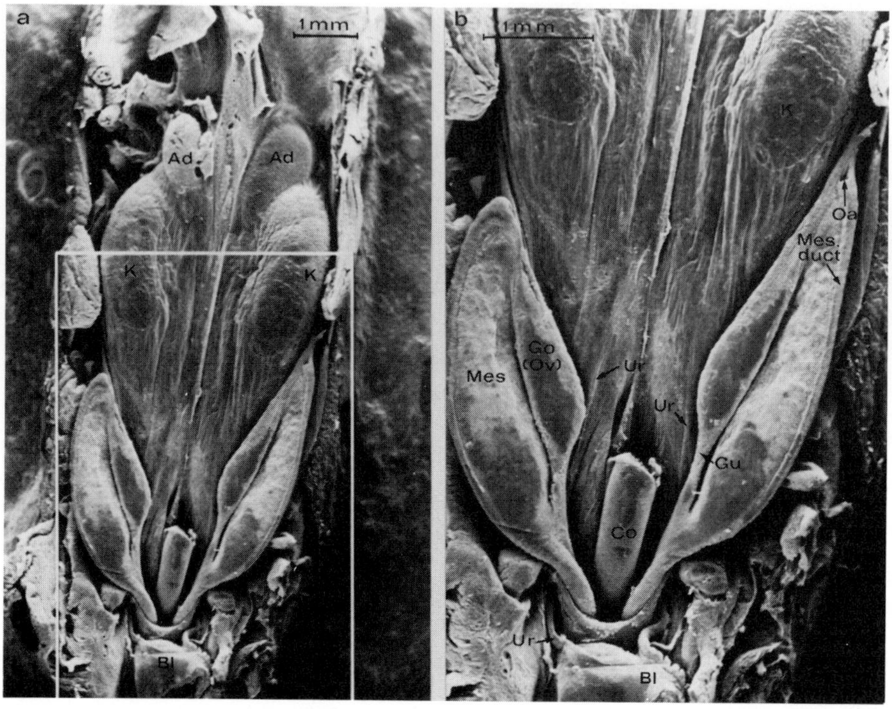

Figure 4: Scanning electron micrograph of the internal genitalia of a 10-day-old female tammar pouch young. (a) Low power magnification of abdominal cavity (b) Higher power magnification of inset shown in (a). Abbreviations as for Figure 3 and Ov, ovary; Ur, ureter.

anlagen, and we agree with this conclusion.

External genitalia: In both sexes there is a prominent genital tubercle at birth, which grows only slowly relative to the rest of the body. The penis of the male does not become morphologically distinguishable until after day 65 of pouch life.

SPONTANEOUSLY OCCURRING INTERSEXES

As is the case with eutherians, we can get some valuable clues about marsupial sexual differentiation by studying those experiments of Nature, the spontaneously occurring intersexes.

Sharman et al[24] have described sexual differentiation in four intersex marsupials of known karyotype, and the results are particularly informative. One tammar wallaby was Klinefelter-like, in that it had a 17,XXY karyotype (normal: 2n=16). However, the animal had a female phenotype, with a pouch that contained four well-developed teats and mammary glands, and no scrotum. Internally, there were intra-abdominal testes with no evidence of spermatogenesis, a normal epididymis, vas deferens, prostate and penis, and no gross or histological evidence of derivatives of the Mullerian duct. Since the intra-abdominal testes had obviously succeeded in masculinizing the whole of the internal genitalia and inhibiting development of the Mullerian duct, it is curious that there was no scrotal development, and yet there was persistence of the pouch, nipples and mammary glands (see Figure 5).

Another tammar wallaby was Turner-like, with a 15, XO karyotype. Although the phenotype was basically female, the animal had a pouch with two small teats and rudimentary mammary tissue on one side only, and a small scrotum apparently on the opposite side. Internally, the gonads were in the ovarian position, and contained both ovarian and testicular tissue, but the internal genitalia were indistinguishable from those of a normal female. Since there was no evidence of Wolffian duct stimulation internally, it is remarkable that there was a scrotum (see Figure 6).

An intersex euro (*Macropus robustus*) (see Figure 5) and a brush-tailed possum (*Trichosusus vulpecula*) both had normal male karyotypes and normal male reproductive tracts, with the exception that the testes were intra-abdominal. However, externally, neither animal had a scrotum, but both had a pouch that contained the normal number of teats and associated mammary tissue. Thus, once again, we have a situation in which the testes had successfully masculinized all the internal genitalia, but had failed to induce scrotal development, or inhibit pouch or mammary development.

Recently, an intersex Tasmanian devil, *Sarcophilus harrisii*, has been described (R.L. Hughes & G.B. Sharman, *pers. comm.*) with a mixed XXX, XX, XO karyotype and half a pouch, two teats and mammary glands on the right side of the body, and half a scrotum on the other side. Internally, the genitalia were essentially female, and ovaries were present. How could an animal be a bilateral gynandromorph externally, with a scrotum on one side and a pouch, teats, and mammary glands on the other side, when the internal organs show no evidence of androgenic stimulation?

HORMONAL MANIPULATIONS

A number of early workers, most notably Burns and his colleagues, recognized the value of the opossum's undifferentiated state at birth, and attempted to modify the development of the sex organs experimentally by application of sex hormones. Unfortunately, most of these studies were conducted with massive doses of hormones of questionable purity, so that interpretation of their results is somewhat difficult. However, a number of interesting observations were made, and more recent authors using lower doses of pure steroids have confirmed some of the earlier results.

SEXUAL DIFFERENTIATION IN MARSUPIALS

Virginia Opossum

Many of the earlier experimental studies of hormonal effects on sexual differentiation in opossums have been reviewed in detail by Burns[14], so the results will only be summarized briefly here.

Gonads: Development of the gonads in most cases was unaffected by neonatal treatment with testosterone or estradiol, despite the application of massive (often lethal) amounts of hormone. However, Burns observed which male opossums which were repeatedly treated with 0.2-0.3 g estradiol dipropionate every 2-3 days from birth, for 20-30 days, underwent varying degrees of gonadal sex reversal, depending on the dose and duration of treatment[13,14]. In some cases the reversal was so complete that the resultant ovotestis was almost indistinguishable from a normal ovary. Treatment of male young which was initiated several days after birth was without effect.

This remarkable result has recently been confirmed by Fadem and Tesoriero[15] using the grey opossum, Monodelphis domestica. They found that treatment of genetic male young, on day 1 and 3, with estradiol led to complete suppression of testicular development, as determined at the time of autopsy 22 weeks later. In these estrogenized males, the internal and external genitalia were completely female, although the animals retained their scrota. Treatment of neonatal male or female Monodelphis with testosterone, or treatment of females with estradiol, was apparently without effect on gonadal or genital differentiation.

Wolffian duct: Treatment of Didelphis pouch young with testosterone led to retention of the Wolffian duct in females, and development of an epididymis[14]. Paradoxically, treatment with estradiol also induced hypertrophy of the Wolffian duct in both sexes.

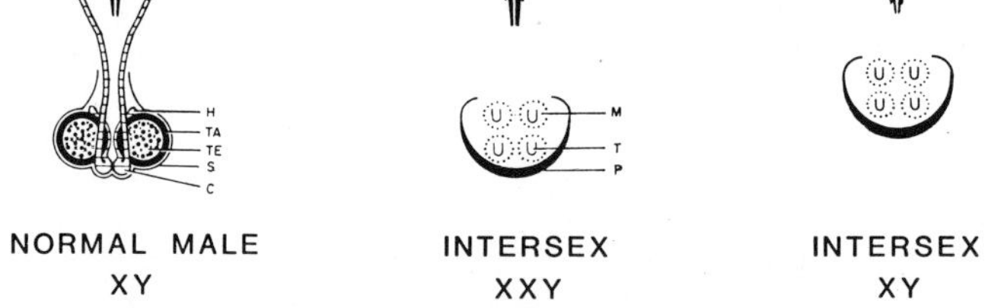

Figure 5: Anatomical development of normal male and intersex macropodid marsupials. Note that the XXY intersex tammar has normal male internal genitalia, intra-abdominal testes, no scrotum, but a normal pouch which contains four teats and four mammary glands. The XY intersex euro is also very similar. V, vas deferens; Pu, Prostatic urethra; PE, penis; H, caput epididymis; TA, tunica albuginea; TE, testis; S, scrotum; C, Cauda epididymis; B, bulbo-urethral gland; M, mammary gland; T, teat; P, pouch. Adapted from Sharman et al., 1970[24].

Mullerian duct: Treatment of Didelphis with large doses of testosterone led to retention of the Mullerian duct in both sexes, although Mullerian duct development was invariably better in females than males[14]. With low doses of testosterone, no retention of the Mullerian duct was seen in male young, even though there was marked stimulation of the Wolffian duct and prostatic development. Treatment with estradiol seemed to accelerate Mullerian duct development in female young.

Urogenital sinus and derivatives: Unlike the gonoducts, which showed paradoxical responses to high doses of male and female sex steroids, the urogenital sinus and derivatives showed the predicted responses in Didelphis. Testoterone exaggerated development in males, with hyperplasia of the sinus epithelium and massive hypertrophy of the prostatic buds[20]. In females, less prostatic hypertrophy was seen[12,21]. Although normal females possess no prostatic rudiments, a dose of 5 μg testosterone per day was sufficient to induce normal male-type prostatic development[10].

On the other hand, estradiol completely suppressed prostatic formation in males, and transformed the sinus epithelium into a stratified squamous type of epithelium, similar to that seen in adult estrous females.

External genitalia: The development of the genital tubercle was affected by exogenous steroids in the predicted manner; testosterone stimulated penile development in the female, and estradiol inhibited penile development in the male[10].

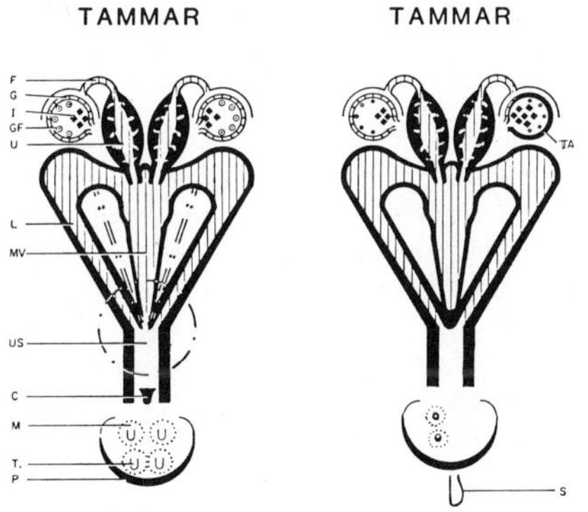

Figure 6: Anatomical development of normal female and intersex tammar wallabies. Note that the XO female has nipples and mammary glands on the left within a pouch, and apparently a hemiscrotum on the right The right gonad has no primordial follicles, but interstitial tissue, seminiferous tubules and a thick tunica albuginea, whereas the left gonad is more ovary-like, although there are no primordial follicles. Adapted from Sharman et al., 1970[24] and Tyndale-Biscoe and Renfree, 1987[25].

It was remarkable that even massive doses of androgen or estrogen had no effect whatsoever on pouch or scrotal development[10].

The brush-tailed possum

Treatment of prepubertal males with massive doses of estrogen caused testicular retraction which resulted from increased tone of the cremaster muscle, and converted the empty scrotum into a pouch-like structure[5,6]. Furthermore, it was claimed that treatment of castrated young possums with estradiol led to the formation of a pouch in the scrotal rudiment[3,4]. The significance of these observations is in doubt, however, since Sharman was unable to reproduce the effect in the quokka *Setonix brachyurus*[23]. Perhaps Bolliger mistook the scrotal invagination which followed the hormonally-induced cryptorchidism or castration for pouch formation.

The tammar wallaby

Daily treatment of a newborn male with estradiol benzoate for 10 days converted the testes into ovotestes[1]. Similar, but less extreme, modifications were seen in two other animals when treatment was initiated on day 2 or 3. The transformation was not as complete as reported in the Virginia opossum, but this observation may reflect the relatively more developed state of the tammar at birth. The estrogen had no observed effect on the Wolffian duct, but it apparently delayed fusion of the Mullerian duct with the Wolffian duct by several days, and caused some hypertrophy of the posterior regions, so that they resembled the lateral vaginae of normal females. Regression of the Mullerian duct was delayed and the *ostium abdominale* became funnel-like, resembling the normal female condition. The treatment with estradiol caused hypertrophy of the urogenital sinus to produce a cornified epithelium, and inhibited phallic development.

SUMMARY AND CONCLUSIONS

The results described here demonstrate that the newborn marsupial is ideal for investigations of the early stages of sexual differentiation. Now that a whole array of pure, synthetically derived hormones and anti-hormones are available, we can confidently expect marsupials to play an increasingly important role in establishing how genetic sex regulates the transformation of the indifferent gonad into a testis or an ovary, and how gonadal hormones subsequently control the development of the Wolffian and Mullerian ducts, and the important processes of trans-abdominal migration and inguinal descent of the testes.

Although the sequence of events in marsupial sexual differentiation is similar to that in eutherian mammals, there appears to be a number of important differences. The results of hormonal administration to pouch young, and a study of spontaneous intersexes, make it clear that androgen is apparently incapable of stimulating scrotal development or inhibiting pouch, nipple or mammary development in females. Likewise estrogen is apparently incapable of inhibiting scrotal development or inducing pouch, nipple or mammary development in males. In eutherians, on the other hand, scrotal development and inhibition of nipple and mammary development are all thought to be androgen-dependent events. If it can be confirmed that development of the marsupial nipple, pouch, mammary gland and scrotum is not under hormonal control, we will have to consider the intriguing possibility that their sex-specific differentiation is genetically determined.

It is also becoming increasingly clear that Burns[13,14] was probably right in his assertion that low doses of estrogen can cause partial transformation of the indifferent gonads of male Virginia opossums into ovaries. Alcorn[1] has obtained a similar result in a tammar wallaby, and Fadem and Tesoriero[15] have shown that estrogen can completely inhibit testicular development in the grey opossum. These effects are probably due to a direct action of estrogen on the development of the indifferent gonad of the male, and even though the effects may be pharmacological rather than physiological, they provide a novel experimental tool for studying the testis-inducing action of the Y chromosome. It is interesting that neither androgen nor estrogen has any effect on the indifferent gonads of female marsupials. The sex-reversed gonads of male marsupials deserve further study, especially with respect to the fate of the germ cells.

Finally, there is the age-old question about the embryological origin of the marsupial's pouch. Pouched marsupials are thought to have evolved from pouchless forms, so it would appear that the pouch is of more recent origin than the scrotum. This hypothesis makes it improbable that the pouch and scrotum have evolved as strictly homologous structures, unlike the labia major and scrotum of eutherians. It seems safer to conclude that the pouch and scrotum may share a partial homology, arising from different parts of the same anlagen. While both structures may be present simultaneously in the pouch young of some species, such as <u>Didelphis</u>, in adult animals and even in intersexes they appear to be mutually exclusive. Until we understand more about the factors that control pouch and scrotal development, we will not be able to solve this intriguing problem.

REFERENCES

1. Alcorn, GT: Development of the ovary and urino-genital ducts in the tammar wallaby <u>Macropus eugenii</u> (Desmarest, 1817) Ph.D. thesis <u>Macquarie University, Sydney, 1975.</u>

2. Alcorn, GT, Robinson, ES: Germ cell development in female pouch young of the tammar wallaby <u>Macropus eugenii</u>. <u>J Reprod and Fertil</u>, 67:319-25, 1983.

3. Bolliger, A: Functional relations between scrotum and pouch and the experimental production of a pouch-like structure in the male of <u>Trichosurus vulpecula</u>. <u>Journal and Proceedings of the Royal Society of New South Wales</u>, 76:283-293, 1943.

4. Bolliger, A: An experiment on the complete transformation of the scrotum into a marsupial pouch in <u>Trichosurus vulpecula</u>. <u>Medical Journal of Australia</u>, 2:56-58, 1944.

5. Bolliger, A, Carrodus, A: Experimental testicular ascent as produced in a marsupial (<u>Trichosurus vulpecula</u>). <u>Australian and New Zealand Journal of Surgery</u>, 9:155-163, 1939.

6. Bolliger, A, Canny, AJ: The effect of the synthetic oestrogens, stilboestrol and hexoestrol on the pouch and scrotum of *Trichosurus vulpecula*. *Journal and Proceedings of the Royal Society of New South Wales*, 75:21-36, 1941.

7. Burns, RK: The differentiation of sex in the opossum (*Didelphys virginiana*) and its modification by the male hormone testosterone propionate. *J Morph*, 65:79-199, 1939a.

8. Burns, RK: Sex differentiation during the early pouch stages of the opossum (*Didelphys virginiana*) and a comparison of the anatomical changes induced by male and female sex hormones. *J Morph*, 65:497-547, 1939b.

9. Burns, RK: The origin of the rete aparatus in the opossum. *Science*, 94:142-144, 1941.

10. Burns, RK: Hormones and experimental modification of sex in the opossum. *Biological Symposia*, 9:125-146, 1942.

11. Burns, RK: Bisexual differentiation of the sex ducts in opossums as a result of treatment with androgen. *J Exper Zoology*, 100:119-140, 1945.

12. Burns, RK: Hormones versus constitutional factors in the growth of embryonic sex primordia in the opossum. *Amer J Anat*, 98:35-67, 1956a.

13. Burns, RK: Transformation du testicule embryonaire de l'opossum en ovotestis ou en "ovaire" sous l'action de l'hormone femelle, le dipropionate d'oestradiol. *Archives d'Anatomie Microscopique et de Morphologie Experimentale*, 45:173-202, 1956b

14. Burns, RK: Role of hormones in the differentiation of sex. In *Sex and Internal Secretions*, 1:76-158, ed. W.C. Young, Baltimore, Williams and Wilkins, 1961.

15. Fadem, BH, Tesoriero, JV: Inhibition of testicular development and feminization of the male genitalia by neonatal estrogen treatment in a marsupial. *Biol of Reprod*, 34:771-776, 1986.

16. Finkel, MP: The relation of sex hormones to pigmentation and to testis descent in the opossum and ground squirrel. *Amer J Anat*, 76:93-152, 1945.

17. George, FW, Hodgins, MB, Wilson, JD: The synthesis and metabolism of gonadal steroids in pouch young of the opossum, *Didelphis virginiana*. *Endocrinology*, 116:1145-1150, 1985.

18. Hayman, DL, Martin, PG: Mammalia I: Monotremata and Marsupialia. In *Animal Cytogenetics, Vol. 4, Chordata 4*, pp 1-110, ed. B. John, Berlin: Gebruder Borntraeger.

19. McCrady, E: The embryology of the opossum. *American Anatomical Memoirs*, 16:1-233, 1938.

20. Moore, CR: Modification of sexual development in the opossum by sex hormones. Proceedings of the Society for Experimental Biology, 40:544-546, 1939.

21. Moore, CR: Prostate gland induction in the female opossum by hormones and the capacity of the gland for development. Amer J of Anat, 76:1-31, 1945.

22. Morgan, CF: The normal development of the ovary of the opossum from birth to maturity and its reactions to sex hormones. J Morph, 73:27-85, 1943.

23. Sharman, GB: Marsupial reproduction. In Biogeography and Ecology in Australia. Monographiae Biologicae, 8:332-368, 1959.

24. Sharman, GB, Robinson, ES, Walton, SM, Berger, PJ: Sex chromosomes and reproductive anatomy of some intersexual marsupials. J Reprod and Fertil, 21:57-68, 1970.

25. Tyndale-Biscoe, CH, Renfree, MB: Reproductive physiology of marsupials. Cambridge University Press, 1987.

26. Ullman, SL: The origin of the granulosa cells in bandicoots. Proceedings of the Australian Society for Reproductive Biology 18:131, 1986.

ANTIQUITY OF THE GENETIC MECHANISM OF SEX DETERMINATION AND THE IMMORTALITY OF THE PRIMARY SEX DETERMINING GENE

Susumu Ohno

Beckman Research Institute of the City of Hope
1450 East Duarte Road
Duarte, CA 91010

In his last masterpiece "ADA", Vladimir Nabokov left us with these immortal words of truth: "Whereas ordinary mortals are content to immitate others, creative geniuses are condemned to plagiarize themselves". Indeed, the paucity of true innovations is also the hallmark of evolutionary processes. In 1970, I wrote the monograph entitled "Evolution by Gene Duplication" without fully realizing its implications[1] The mechanism of gene duplication provides nothing but the means to plagiarize extensively merely a few innovations. Nearly all the components of the adaptive immune system which is peculiar to vertebrates have been built by plagiarism of a single ancestral domain: the so-called β_2-microglobulin-like domain, made up of roughly 100 amino acid residues, folded to form a series of antiparallel β-sheet loops compacted by one disulfide bridge (Figure 1). It seems as though nature was at its innovative best at the very beginning of life on this earth, some 3.5 or more billion years ago. For example, most, if not all, of the sugar-metabolizing enzymes (various kinases as well as dehydrogenases) apparently acquired their functional competence before the division of eukaryotes from prokaryotes: homologous enzymes of bacteria and mammals still share amino acid sequence homologies of around 50%. Similarly, hamster β-adrenergic receptor has been found to be homologous with bovine retinal opsin. These two similar proteins, with very divergent functions, are both characterized by seven hydrophobic, membrane-spanning helices. These helices, each 20-25 residues long, are repeats of each other. Inasmuch as the basic architecture noted above is also found in bacteriorhodopsin, it would appear that the ultimate ancestor for this family of genes was also in existence before the division of eukaryotes from prokaryotes[2].

The ultimate origin of hemoglobin genes is also of extreme interest. In vertebrates, hemoglobins are encased in circulating erythrocytes, and the genomes of certain teleost fish and more advanced organisms contain two sets of genes; one set for the α-chain and its allies and the other for the β-chain and its allies. Within vertebrates, hemoglobin polypeptide chains have been changing rather rapidly, with a divergence of 1% in the amino acid sequence every 8.3 million years. By contrast, glyceraldehyde 3-phosphate dehydrogenase, one of the sugar-metabolizing enzymes, has been undergoing a 1% change in its amino acid sequence every 40 million years. Reflecting the rapid evolutionary changes, monomeric hemoglobins of lampreys are seen to be already intermediate between myoglobins on one hand and α- and β-chains of hemoglobins of jawed vertebrates on the other[3]. Thus, within vertebrates, all the indications are that the gene duplication event, which yielded the ancestral hemoglobin gene from a redundant copy of the myoglobin gene, must have taken place at the onset of vertebrate evolution, some 300 million years ago. However, it has been known for a long time that hemoglobins appear sporadically, not only among invertebrates (e.g., in Chrinoromus among dipteran insects, in earthworms among the class Polychaeta of the phylum Annelida) but also among the plants (e.g., in nitrogen-fixing nodules of leguminous plants). A dimeric bacterial hemoglobin from Vitreoscilla has recently been sequenced[4]. It is comprised of 146 amino acid residues, and it is the same length as mammalian -chains. Furthermore, all the functionally critical residues are present., e.g. a pair of histidine residues that hold a heme group, and phenylalanine at position 46, which is invariant in all hemoglobins. This bacterial hemoglobin shows the greatest sequence homology (24%) with pea leghemoglobin which is 153 residues long. The fascinating evolutionary history of hemoglobins confirms the view that most of the major innovations in evolution occured at the very beginning of life on this earth, before the division of eukaryotes from prokaryotes. In addition, it reveals yet another evolutionary principle which is often overlooked[5]. A gene, once invented, may remain dormant for a very, very long time, only to be resurrected in certain organisms. For example, insects, as a rule, do not express hemoglobin genes. Even among dipteran insects, the familiar Drosophila and mosquitos do not express them, while Chrinoromus does. The gene that can be resurrected after a very long period of dormancy must necessarily be endowed with the immortal property of being impervious to deleterious mutations that cause premature chain termination, reading frame shifts etc., and that can deprive the gene product of its function[5].

In this paper, we shall discuss the antiquity of the sex determining mechanism within the context of the hypothesis that the major, primary, sex determining genes are immortal. In various phylogenetic trees, the sex determining mechanism disappeared once, only to be resurrected later on.

ANTIQUITY OF THE MECHANISM FOR SEX DETERMINATION

As with all the ultimately primordial genes, the primordial genes for the sex determining mechanism seem also to have evolved shortly after the creation of life on this earth, not before the division of eukaryotes from prokaryotes, but after unicellular eukaryotes acquired the diploid set of chromosomes. A priori, one is tempted to consider the "invention" of meiosis and syngamy as a pair of events that occurred together. It is possible, however, that syngamy developed first; the transient tetraploidy may have preceeded the segregation of diploid sets.

Mating types of the baker's yeast, Saccharomyces cerevisiae, are very well known. Furthermore, among unicellular organisms, it has recently been shown that African trypanosomes recombine with each other while residing in their intermediate host, the tsetse fly[6].

In the case of the multicellular eukaryotes, sex chromosomes, as such, apparently evolved before the division of plants from animals. In the plant kingdom, sex chromosomes, which are as conspicuously heteromorphic as mammalian sex chromosomes, are commonly found among mosses

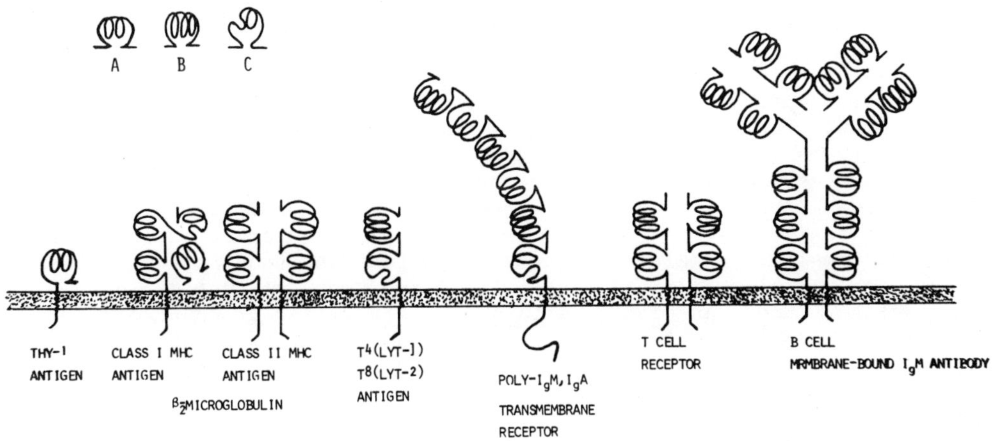

Figure 1: A schematic demonstration of the main components of the adaptive immune system as they arose from extensive plagiarism of a simple ancestral gene which encodes a 90- to 100-residue-long polypeptide chain, folded to form two β-sheet structures held together by one intradomain disulfide bridge. At the top left, A signifies the unit type with underdeveloped 2nd β-sheet structure, which is probably the ultimate ancestor; a direct descendent of portions of N-CAM-like, cell-adhesion proteins. B signifies the derived unit type which gained a tryptophan loop in the second β-sheet structure. This innovation made the formation of the dimeric, antigen-binding pocket possible. C signifies the degenerate unit type, although the definition of C largely depends upon individual choices. At the bottom, aligned from left to right on the plasma membrane, which is indicated by the shaded portion, are: 1) Thy-1 antigen (A-type monomer), 2) Class I MHC antigen (C + A + A trimer) polymerized with β_2-microglobulin (A-type monomer), 3) Class II MHC antigen (both α- and β-chains being A + A dimers), 4) T4, T8 antigens (B + C dimers), 5) Poly-IgM, IgA transmembrane receptor (B + A + B + A + B + C hexamers), 6) T-cell receptor (both α- and β-chains being B + A dimers), 7) Membrane-bound IgM (light chain is B + A dimer, while μ-class heavy chain is B + A + A + A + A pentamer).

and liverworts of the phylum Bryophyta. Since the only plants simpler than bryophytes are the algae and fungi, it is likely that primordial, multicellular plants were, indeed, endowed with the sex determining mechanism; the true hermaphroditism seen in most of the modern flowering plants probably represents a secondary type of development. In the animal kingdom, the sex determining mechanism operates in the majority of species. However, when we confine ourselves to consideration of vertebrates, it seems as though the chromosomal sex determining mechanism evolved anew in advanced vertebrates, and that the original vertebrates were hermaphrodites of one kind or another. This idea, however, is an illusion. It is granted that a conspicuously heteromorphic pair of sex chromosomes is not to be found among fish. Nevertheless, the great majority of fish, from sharks to teleosts, is gonochoristic and produces males and females in roughly equal proportions[7]. It is true that there are a number of hermaphroditic species of teleost fish of various types: 1) synchronous hermaphrodites, possessing a functional ovary and a testis or a pair of ovotestes at the same time; 2) asynchronous hermaphrodites of two types: protogynous (females when young and males as they become older) and protoandrous (young males transforming into females with age). Perhaps the most fascinating of the protogynous hermaphrodites are certain marine fish inhabiting tropical coral reefs. In the case of Anthias sqamipinnis, found around Puerto Rico, the removal of an existing male from a bisexual social group triggers the transformation of the dominant female to a male. The cues triggering the onset are not simply visual, acoustic or chemical, involving close contact and behavioral interaction. Since Anthias squamipinnis is a rather pronouncedly sexually dimorphic species, the female-to-male transformation involves not only the conversion of ovaries to testes but also conspicuous alterations in external coloration[8].

As fascinating as these hermaphorditic species are, they represent secondary modifications, developed after the tremendous expansion and adaptive radiation of teleost fish that began rather recently in the Cretaceous period, a mere 130 million years ago. Hermaphroditism is not found among sharks and rays of the subclass Elasmobranchin; nor among members of the superorders Chondrostei and Holostei; nor among surviving members of the choanichthian lineage that gave rise to terrestrial vertebrates; nor in lungfish of the order Dipnoi and more pertinently in the Latimeria (coelacanth) of the order Crossopterygii. The coelacanth incidentally, is ovoviviparous. Thus, it would appear that the ancestral vertebrate of the Ordovician era, some 400 million years ago, was also a gonochorist, endowed with the sex determining mechanism. However, sessile filter-feeders of early Cambrian times, resembling the tunicates of today, may have been true hermaphrodites. It should be recalled that the basic body form of vertebrates is embodied in the free-swimming, tadpole-like larvae of these sessile filter feeders. In evolution, a process called paedomorphosis happened time and time again. The adult stage was eliminated and the larval form became sexually mature and reproduced. I would venture a guess that it was during such paedomorphosis that the sex determining mechanism was regained.

Among reptiles, temperature-dependent sex determination was first found in turtles and tortoises[9,10]. Subsequently, the same mechanism was found to operate in another group of ancient reptiles, the aligators and crocodiles. The existence of this mechanism has often been interpreted, or rather misinterpreted, to mean that ancient reptiles had no genetic mechanism of sex determination. It should be recalled, however, that these reptilian eggs, when hatched at the optimum temperature, yield males and females in nearly equal numbers. Only at extreme temperatures do either all male or all female progeny result. It would appear, therefore, that the observed dependence on temperature is merely a reflection of the acquired temperature sensitivity of a preexistent genetic mechanism of sex determination[11].

As is the case for all the major innovations in evolution that occurred during the existence of life on this earth, the sex determining mechanism appears also to have established itself soon after the beginning of life, 3.5 or more billion years ago. Unicellular eukaryotes have such a mechanism, and so did ancestral plants. Inasmuch as all these initial, major innovations have since been plagiarized to the extreme, with no additional, true innovations in divergent phylogenetic trees, it is likely that components of the sex determining mechanisms have remained essentially unchanged since the beginning.

THE SAME SET OF GONAD-FORMING GENES IS REQUIRED BY BOTH GONOCHORISTIC AND HERMAPHORDITIC SPECIES

All the marvelously diverse sexual phenotypes seen within the same order, or even within the same family, are just manifestations of the variable subsidiary system that is only secondarily associated with the primary sex determining mechanism of great antiquity. For example, among deer of the family *Cervidae* of the order *Artiodactyla*, antlers adorn the heads of stags but not of does of the majority of species, e.g. the old-world red deer *Cervus elephus* and the new-world elk *Cervus canadiensis*. In these predominant species, the growth of antlers is an androgen-dependent trait. In the diminutive musk deer *Moschus moschiferus*, however, even males are antlerless, while their abdominal subcutaneous glands supply precious substances for expensive perfumes. The growth of antlers has become androgen-noninducible in this species. In several reindeer species of the genus *Rangifer*, on the other hand, antlers are grown by both males and females alike. In this genus, the growth of antlers has become a consitutive trait no longer requiring induction by androgenic steroids.

In sharp contrast to the dependence or independence of antlers on androgenic steroids, the primary sex determining mechanism has been doing the same task, with monotonous invariance, since the establishment of the first metazoans, namely, construction of an ovary and/or a testis. For so simple a task, not many genes are needed. In fact, G.M. Edelman has recently made a rather convincing argument that at the initial stages of embryonic development, the onset of organogenesis is probably controlled by a small family of CAM (Cell Adhesion Molecule) proteins, e.g., C-CAM and N-CAM[12]. The backbones of these CAM glyco-proteins are constructed from a series of β_2-microglobulin-like domains. Therefore, they might be immediate ancestors of many components of the adaptive immune system[13]. These CAM proteins, anchored on the plasma membrane, are involved in homologous recognition, and cause the aggregation of cells that express the same CAM. We have argued similarly that organization of an ovary and/or a testis should involve a small number of plasma membrane proteins[14]. In essence, various organs can be classified into a small number of types, of which one is the glandular type. Aside from organs identified as glands (salivary, thyroid, etc.), the pancreas and kidneys, among others, belong to this type. The testis, comprised of seminiferous tubules, is obviously of the glandular type. While it is not very obvious in ovarian development of small rodents with very short gestation periods, the early ovary of all large mammals, including man, develops very conspicuous ovigerous cords which are counterparts of early seminiferous cords[15]. These cords are broken up only when individual oocytes within them begin to organize primordial follicles. Thus, it is likely that the set of plasma membrane proteins involved in the organogenesis of ovaries and testes belongs to the same subfamily as the set involved in organogenesis of other organs of the glandular type. The minimal requirement for gonadal organogenesis is one pair of proteins, both anchored to the plasma membrane: the ovary organizer (O-CAM) which, by homologous recognition, organizes the ovary; and the testis organizer (T-CAM) which engages in the complementary recognition with O-CAM (Figure 2). Both O-CAM and T-CAM, and especially T-CAM, may detach from the plasma membrane to function as short-range,

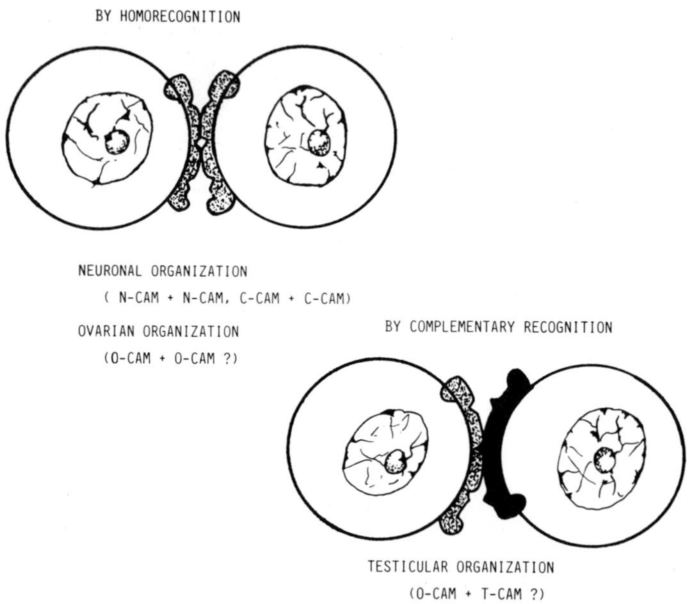

Figure 2: The role of cell adhesion molecules (CAM) in cell-cell contacts and, therefore, in organogenesis. Organization of amoeboid slime molds into stalk and fruiting body by aggregation, as well as the neuronal organization of the chicken, is mediated by homologous recognition between molecules of the same plasma membrane proteins. In the case of neuronal organization, two proteins are involved: a more ubiquitously expressed C-CAM and a neuronal cell-specific N-CAM. When gonadal organogenesis is viewed from a similar vantage point, the minimal requirement is the presence of two CAM's. The ovary might be organized by homologous recognition between O-CAM's, drawn as shaded, elongated bodies on the plasma membrane. If such were the case, testicular organogenesis might involve the complementary recognition between O-CAM and T-CAM, drawn as the solid, elongated body on the plasma membrane. Complementary recognition is expected to be characterized by a higher binding affinity than homologous recognition, so that in this scheme testicular organogenesis should enjoy a dominance over ovarian organogenesis. Two CAM's engaged in complementary recognition can belong to the same family and, therefore, share a common ancestor. Indeed, major histocompatibility antigens and the receptors on T-cells, designed to recognize them by complementarity, belong to the same β_2-microglobulin family.

paracrine hormones[16]. It should be recalled that the majority of experimentally produced XX/XY chimeric mice manages to organize a pair of testes[17]. The recognition based upon complementarity is expected to exercize dominance over homologous recognition. It is only in synchronously hermaphroditic species that this pair of genes must be expressed simultaneously at a certain stage in embryonic development and possibly thereafter. But even in this case, one or other of the genes should be regulated differentially:

1) In the case of lateral, true hermaphrodites, the O-CAM gene may be expressed in both indifferent gonads, but T-CAM should be expressed only on the right or on the left.

2) In the case of ovotestis-bearing, true hermaphrodites, O-CAM may be expressed in all parts of both indifferent gonads, but the expression of T-CAM should be confined to the meduallary portion of each gonad.

3) In the case of asynchronous hermaphrodites of both the protogynous and protoandrous types, the sequential regulation of expression of O-CAM and T-CAM, in each pair of gonads, is required.

The initial expression of O-CAM and T-CAM should be followed by cessation of production of T-CAM, in the case of protoandrous species. In the case of protogymous species, the age-dependent induction of T-CAM must necessarily occur. Thus, it is a mistake to believe that development of the genetic mechanism for sex determination in gonochoristic species represents an innovation that requires the addition of a new regulatory gene or genes. Gonadal development was meant to be differentially regulated under all circumstances. Furthermore, even among gonochoristic species, both males and females of the majority of species apparently possess genes for both the O-CAM and the T-CAM. Even if we confine our discussion to vertebrates, complete sex reversals in both directions (i.e., conversion of genetic males to functional females and of genetic females to functional males) have been accomplished in one gonochoristic species of teleost, *Oryzias latipes*[18], as well as in one gonochoristic species of frog, *Xenopus laevis*[19]. These complete sex reversals were accomplished by larval administration of sex steroids. The reason that such sex reversals seldom, if ever, occur in nature is that sex steroids are normally produced only after gonadal differentiation by the differentiated gonads themselves. Thus, the situation is no different from that of many other genes with sex-specific expression. For example, although males normally have no use for them, the mammalian male genome is, nevertheless, endowed with all the necessary genes for the development of mammary glands and lactation. Indeed, in certain cases of hormonal imbalance, men may produce substantial amounts of milk, as in cases of galactorrhea. While the pronouncedly heteromorphic pair of sex chromosomes, XY of the male heterogamety as well as the ZW of the female heterogamety, is seldom found in fish, amphibians and reptiles, the ZW pair of a majority of the avian species is as heteromorphic as the mammalian XY pair. However, heterogametic females of these avian species are endowed with only one functional ovary, on the left side. Early removal of this left ovary induces compensatory hypertrophy of the right residual gonad which invariably develop into a testis, in which complete spermatogenesis and spermiogenesis may take place[20]. Inasmuch as the above noted sex reversal of avian genetic females to potentially functional males reveals that the genome of the heterogametic female is endowed with genes for both O-CAM and T-CAM, the same must necessarily hold true for the homogametic male sex: the homogametic sex can have no gene that is uniquely its own. The sex determining mechanism of avian species may depend upon the dosage effect of T-CAM genes on the Z chromosome.

At present, it appears as though complete sex reversals are an impossible proposition only in placental mammals. Inherited sex reversals, such as XX males and XY females should be excluded, for such reversals involve mutational changes in the genes for gonadal organogenesis themselves. If, indeed, the testis of placental mammals can be organized only in the presence of a normally Y-linked gene or genes, the situation in such mammals is indeed unique. The simplest explanation might be that the T-CAM gene resides on the mammalian Y chromosome. Implicit in this assumption is the constitutive expression of the O-CAM gene by females and males alike. Such common expression seems to pose no problem, however, for it should be recalled that in XX/XY chimeras, a minority (as few as 20%) of XY cells is capable of enticing neighboring XX cells in the gonad to engage in testicular organogenesis, as already noted. This result is to be expected, since the complementary recognition between O-CAM and T-CAM should have a higher binding affinity than the homologous recognition between molecules of O-CAM. How can we account for those nonmammalian species, in which the Y of the male heterogamety, as well the W of the female heterogamety, is of no obvious consequence. In fact, the W chromosome of avian species might be of this inconsequential type. The majority of 3AZZW triploid chickens develop as males in spite of the presence of the W chromosome[16]. As already noted, in avian species, the T-CAM gene may reside on the Z chromosome, two doses of it sufficing for successful testicular organogenesis. In the presence of only one dose of T-CAM gene, as in ZW females, the homo-recognition between molecules of O-CAM might take over, thereby organizing a functional ovary on the left side. Nevertheless, the right residual gonad is organized in the testicular direction, so that once the inhibition exerted by the left functional ovary is removed, it can hypertrophy to become a true testis.

It would, thus, appear that the difference between hermaphroditic species and gonochoristic species is a very minor one, since differential regulation of expression of O-CAM and T-CAM (e.g. the homo-recognition between molecules of O-CAM for ovarian organogenesis, and the higher affinity, complementary recognition between molecules of O-CAM and T-CAM for testicular organogenesis) is required by all species. In fact, the differential regulation which operates in hermaphroditic species has to be far more sophisticated and precise than that which operates in gonochoristic species: i.e. the organization of a testis on one side and an ovary on the other side of the same individual requires a far more finely tuned mechanism for regulation than the generation of either a pair of testes or a pair of ovaries in each individual.

THE ULTIMATE ANCESTRAL GENE FOR CAM (CELL ADHESION MOLECULE) PROTEINS

In view of nature's inherent inclination towards plagiarism of a few, true innovations, it is rather to be expected that plasma membrane proteins, involved in ovarian and testicular organogenesis, belong to the same family or superfamily of proteins which includes C-CAM and N-CAM. With the exception of an amino terminal portion of unknown length, the amino acid sequence of N-CAM of the chicken has recently been deduced from the nucleotide sequenceof cDNA[13]. Its extra cellular portion consists of four successive β_2-microglobulin-like domains to which the near-membrane domain, 200 residues long, is attached. Below the transmembrane region, comprised of 18 hydrophobic residues, there exists the 245 residues long intracellular domain. While it is clear that the β_2-microglobulin-like domains of N-CAM, and possibly also of C-CAM and other cell adhesion molecules required for organogensis, served as immediate ancestors of various components of the vertebrate, adaptive immune system depicted in Figure 1, from where did these β_2-microglobulin-like domains of N-CAM and other CAM's arise? I believe that they are but modified versions of the rest of the CAM. Aside from two cysteine residues, spaced roughly 40 residues apart, which invariably form one intradomain disulfide bridge, each β_2-microglobulin-like domain is characterized by its hydrophobicity, which results from an abundance of Val, Leu, Ile, Phe and Tyr residues. It should be recalled that without glycosylation, immunoglobulin molecules are nearly insoluble in water. The relative abundance of Ser and Thr residues is another characteristic of functional importance, for neighboring β-sheet-forming strands are held in close proximity to each other by hydrogen bonds formed mostly between Thr and Thr, Thr and Ser or Ser and Ser residues. Of the various domains of N-CAM, the intracellular domain (from the 688th residue to 932nd residue in the nomenclature of Hemperly[13]) appears to be of the simplest construction, thus, appearing primordial. As seen in the beginning portion of the intracellular domain shown in Figure 3, this domain appears to be of the simplest construction, and to have evolved from the simple dipeptidic repeats Thr-X, Thr-X. An indication that β_2-microglobulin domains of N-CAM evolved from permutations of this simply constructed, intracellular domain is also shown at the top of Figure 3. Thr-X, Thr-X repeats also occur in these domains, particularly in the vicinity of all the important cysteine residues already mentioned and the invariably present Trp residue that occupies the 24th position of each domain.

This recurrence of Thr X, Thr X dipeptidic repeats provides a link between N-CAM and C-CAM and possibly other CAM of vertebrates and of more primitive creatures. The slime mold Dictyostelium discoideum can be considered to be a link between unicellular eukaryotes and multicellular ones. In the nutrient-rich environment, slime molds exist as unicellular amoeboid creatures. When food becomes scarce, however, they aggregate to form a stalk and fruiting body. This aggregation is again mediated by a cell adhesion protein known as csA. As shown in Figure 4, 20% of this polypeptide chain of 494 residues is comprised of Thr-X, Thr-X dipeptidic repeats[21]. Thus, it would appear that, as with all other ultimate ancestors, the ultimate ancestor of cell adhesion molecules was also composed of the simple repeats[28].

A MUSICAL TRIBUTE TO THE PRIMORDIAL, SEX-RELATED, IMMORTAL GENE

As I stated at the beginning of this paper, nearly all the major innovations in the evolution of genes were accomplished soon after the beginning of life on this earth more than 3.5 billion years ago, before the separation of eukaryotes from prokaryotes. Primordial genes for the primary sex determination appear to have arisen shortly thereafter, coinciding with the birth of unicellular but diploid eukaryotes. Subsequent successes in the tremendously adaptive radiations of living organisms have been due mostly to very extensive plagiarism of these precious, few, initial innovations via the mechanism of gene duplication[1]. Accordingly, an understanding of the prebiotic scenario before the birth of the first cells on this earth becomes absolutely essential.

```
                     117                               122
                     THR   MET / THR   TRP / THR   LYS
2ND IG DOMAIN       A C C A T G/A C G T G G/A C A A A G

                                  199              202
                                  THR   LEU / THR  CYS
                                 A C A C T G/A C C T G T
3RD IG DOMAIN
                     213                                    219
                     THR   TRP / LYS / THR   SER / THR   ARG
                    A C G T G G/A A A/A C T T C C/A C C C G G

                     699                    703
                     THR   THR / GLY / THR  THR
                    A C T A C G/G G C/A C C A C T

                     707                                                     715
                     THR   ILE / THR   GLU / THR   PHE / ALA / THR   ALA
CENTRAL DOMAIN      A C T A T C/A C T G A A/A C T T T T/G C C/A C T G C T

                      720                                                      728
                      THR   SER / GLU / THR   THR / THR   LEU / THR   SER
                     A C G A G C/G A G/A C C A C C/A C C C T G/A C C T C A
```

N-C A M (NEURONAL CELL ADHESION MOLECULE) OF THE CHICKEN

Figure 3: The extracellular portion of N-CAM plasma membrane protein of the chicken is comprised of four successive β_2-microglobulin-like domains, each roughly 90 residues long, and the 200 residue near-membrane domain. The intracellular domain is 245 residues long. The entire polypeptide chain appears to have evolved from simple Thr-X, Thr-X dipeptidic repeats. At the bottom, three successive stretches of Thr-X, Thr-X repeats, found at the beginning of the intracellular domain, here identified as the central domain, are shown. Thr-X, Thr-X dipeptidic repeats are also found in extracellular β_2-microglobulin-like domains, always around the most invariant residues, e.g. the 12th Cys, 24th Trp and 62nd Cys residues of each domain. At the top, such repeats in the 2nd and 3rd of the four domains are identified.

Parts of our galaxy, where stars are still being formed, are thickly populated with molecular clouds which contain large quantities of molecular hydrogen, water, ammonia, carbon monoxide, methyl alcohol, hydrocyanic acid and other simple compounds, and these compounds are chemically reducing in nature rather than oxidizing. These chemicals are the ultimate source of all stars which are formed by gravitational condensation of these molecular clouds[22,23]. The earth is believed to have formed roughly 4.5 billion years ago, and for the first billion years or so, it must have been surrounded by the reducing atmosphere described above, and the ocean, which formed within this early period, no doubt contributed to the start of life. The oldest known sedimentary rock in Greenland has been dated as 3.78 billion years old. While the primeval reducing atmosphere would be very noxious to living organisms once formed, it readily gave rise to amino acids, sugars, purines and pyrimidines[24]. As far as self-replicating nucleic acids (RNA rather than DNA) in the prebiotic world were concerned, what was apparently in short supply were templates of adequate length. When a dehydrated mixture of imidazol,

ATP, GTP, UTP and CTP is dissolved in sea water that contains Zn++, oligonucleotides form: the longest being decanucleotides. In the presence of Zn++, on the other hand, much longer templates (e.g. 100 bases long) can be copied in the proper 3'-5' linkage[25]. The hypercycle theory of Eigen and Schuster[26] is a theory of great conceptual beauty as to how life may have started. However, the theory seems to have been based upon erroneous experiments, previously done by others, which suggested that a primitive nucleic acid polymerase such as Qβ viral polymerase might have been endowed not only with copying function but also with polymerization function. It now appears that the apparent formation of templates of 50 to 60 nucleotides by this enzyme was due to contaminating RNA which was present in the enzyme preparation; the purified Qβ replicase has only the copying function[27]. What good would multiple copies of base oligomers have done for the beginning of life on this earth? However, if a given nucleotide octamer was made up of tandem repeats of a tetrameric unit, say G A U A then base pairing might have occurred between a 2nd unit of the original template G A U A G A U A and the 1st unit of its freshly copied, complementary strand U A U C U A U C as shown below:

```
G A U A G A U A
        . . . .
        . . . .
        C U A U C U A U
```

The next replication would have been prompted by a paired, hydrogen-bonded segment which would have served as a primer, and as a result of the next replication, the originally octameric template would have elongated itself to become dodecameric template. This process of continuous elongation after each round of replication must have been inherent in the construction of repeats. Therefore, the process could have continued indefinitely. The above sequence of events represents one of the number of reasons for believing that the first set of coding sequences were repeats of base oligomers[28]. Since four is not a multiple of three, the complementary sequence of the G A U A repeats, described above, would have given the tetrapeptidic periodicity to a translated polypeptide chain as shown below:

```
 Tyr   Leu   Ser   Ile/   Tyr   Leu   Ser   Ile/   Tyr   Leu   SER   Ile/
U A U C/U A U C/U A U C/ U A U C/U A U C/U A U C/ U A U C/U A U C/U A U C
   Ile / Tyr   Leu   Ser    Ile/ Tyr   Leu   Ser    Ile/ Tyr   leu   Ser
     Ser   Ile/ Tyr   Leu    Ser   Ile/ Tyr   Leu    Ser   Ile/ Tyr
```

This sequence occurs because three consecutive copies of the U A U C tetramer, now 12 bases long, is translated in three different reading frames to become the translational unit that encodes Tyr-Leu-Ser-Ile. By contrast, repeats of a basic hexamer of nucleotides could have given only dipeptidic periodicity to its polypeptide chain. Such was the merit of an oligomeric unit which was indivisable by three. There was yet another, equally important merit. The error rate of non-enzymatic, prebiotic replication of nucleic acid was thought to be of the order of 10^{-2} base pair/replication. Unless the first set of primordial coding sequences was endowed with a measure of inherent immortality, they would soon have degenerated into junk sequences. Provided that the number of bases in the oligomeric unit was not a multiple of three, for example, the U A U C tetramer shown above, primordial coding sequences were indeed endowed with a measure of inherent immortality. Inasmuch as they were designed to encode polypeptide chains of the identical periodicity in

```
          3           6
          THR  ILE / THR  ALA
          A C A A T A/A C A G C T

16                                                                    29
THR  TYR / ILE / THR  ILE / THR  GLY / THR  GLY/ PHE / THR  GLY / THR  PRO
A C A T A T/A T T/A C C A T T/A C A G G T/A C T G G A/T T T/A C A G G A/A C T C C A

85                          92
THR  PRO / PRO / THR  LEU / SER / THR  ILE
A C A C C T/C C A/A C T C T T/T C A/A C A A T A

162               166                173                           179
THR  ILE / VAL / THR  SER            THR  ILE / THR  SER / ILE / THR  PRO
A C A A T T/G T A/A C T T C C        A C C A T T/A C A T C A/A T C/A C C C C A

185                                 193
THR  PRO / THR  ASN / VAL / THR  VAL / THR  GLY
A C A C C A/A C C A A T/G T A/A C C G T A/A C C G G T

210               214                          236          239
THR  GLY / LEU / THR  VAL                      THR  ILE / THR  VAL
A C A G G A/T T G/A C T G T T                  A C T A T A/A C T G T C

260          263                       267                     272
THR  PHE / THR  TYR                    THR  ILE / THR  SER / THR  LYS
A C T T T T/A C A T A C                A C C A T T/A C T T C A/A C A A A A

    283                                  291
    THR  ASN / THR  GLY / THR  ASP / PHE / THR  GLN
    A C C A A T/A C T G G T/A C C G A T/T T C/A C T C A A

295               299                  307          310
THR  MET / GLY / THR  SER              THR  GLY / THR  ASN
A C C A T G/G G A/A C C T C A          A C T G G T/A C T A A T

339          342                       421                425
THR  SER /  THR  LEU                   THR  ILE / LYS / THR  GLU
A C A T C A/ A C T T T A               A C T A T T/A A A/A C T G A A

    427                                  434
    THR  THR / SER / THR  THR / TYR / THR  ILE
    A C A A C C/T C T/A C A A C C/T A T/A C A A T T

437                                     446
THR  PRO / THR  PRO / THR  ASP / THR  ALA / THR  PRO
A C T C C A/A C T C C A/A C T G A T/A C A G C C/A C C C C A

        453                             460
        THR  PRO / THR  GLU / THR  ALA / THR  PRO
        A C T C C A/A C T G A A/A C A G C C/A C C C C A

                          463          466
                          THR  SER / THR  PRO
                          A C C A G C/A C A C C A
```

C A M (Cell adhesion Molecule) OF *Dictyostelium discoideum*

Figure 4: Through the presence of Thr-X, Thr-X dipeptidic repeats, the N-CAM of the chicken can be compared to the earlier CAM which arose at the beginning of multicellular eukaryotes. The CAM of the slime mold, known as csA, clearly evolved from the simple Thr-X, Thr-X repeats. In this figure, 19 stretches of Thr-X, Thr-X dipeptidic repeats are identified. All together, these Thr-X, Thr-X dipeptidic repeats comprise 20% of the entire sequence.

all three reading frames, as shown above, they had to be impervious to most mutational changes that cause shifts in reading frame and premature chain termination, and that deprive proteins of their normal functions.

In fact, DNA repeats of G A T A and its single base deviant G A C A are very ancient in their origin and are somehow associated with the sex determining mechanism from the baker's yeast to mammals[29]. Therefore, I am tempted to believe that such sequences were the ultimate ancestor of the genes for O-CAM and T-CAM. Indeed, not one, but several, transcripts complementary to G A T A, G A C A repeats have been found in the mouse[28]. A portion which contains repeats of U A U C (complementary to G A T A) has just been shown above. In fact, the continued persistence of such a family of primordial coding sequences may indicate that they serve as the reserve which enables those species, which once forfeited the mating system, to regain it at a later date. Analogy to such a possibility might be found in hemoglobins. I have already noted that, aside from their universal occurrence in vertebrates, hemoglobins occur sporadically in invertebrates, plants and even in bacteria. Such sporadic recurrence (so-called "convergent evolution" for whatever it implies) would be impossible without the persistence of an ultimately primordial coding sequence for hemoglobins.

In a previous paper[30], we pointed out the striking similarity between the construction of immortal, primordial coding sequences and that of Renaissance to Baroque musical compositions. Modern coding sequences, 3 billion or more years removed from their ultimate ancestors, on the other hand, are constructed along the lines of late Romantic musical compositions. Thus, we devised a set of rules, whereby treble clef musical scores and base sequences become interchangeable. Inasmuch as the octave scale has to accomodate only four bases, a space and a line are assigned to each base with the ascending order of A, G, T, C. The C of the ascending order occupies the classical middle C position. The resulting redundancy introduced to the positioning of bases is quite analogous to the redundancy of codons. Emerging life in the prebiotic world faced the need to encode 20 amino acids with combinations of four bases. Inevitably, the coding system based on purine, pyrimidine triplets emerged. Accordingly, in the universal coding system, only two amino acids, Met and Trp, are encoded by one codon each, others are specified by two, three, four and even six codons each. Consequently, a very large variety of base sequences can encode the identical amino acid sequence. Hence, those researchers who wish to clone a particular coding sequence, which is based upon a sequenced peptide fragment, have to prepare not one but a variety of synthetic oligonucleotide probes. Similarly, a given coding sequence under our set of musical rules can be transformed into not one but a number of musical scores. The number, however, is not very large, if one is to avoid unpleasant discordance.

However, the converse in both the biological and musical context is without ambiguity. One coding sequence in a given reading frame can only be translated into one amino acid sequence. Similarly, a given treble clef musical score can only be transcribed back to one unique, particular base sequence. In Figure 5, a portion of the mouse transcript complementary to G A T A, G A C A repeats is transformed to the musical score in the manner of J.S. Bach, (i.e. his prelude No. 1 for the well-tempered clavichord).

SUMMARY

All the available evidence indicates that most, if not all, of the ultimately primordial genes arose shortly after the emergence of life on this earth, before the separation of eukaryotes from prokaryotes, some 3.5 billion or more years ago. Subsequent adaptive radiation of life on this earth, on a tremendous scale, has been accomplished largely by extensive plagiarism, via the mechanism of gene duplication of these precious few, initial innovations. I attribute this initial innovativeness to the construction of primordial coding sequences which were repeats of

MOUSE GATA GACA SEX RELATED TRANSCRIPT

Figure 5: The musical transformation in A minor and 8/8th time signature ← of a portion of the mouse transcript complementary to the primordial sex-related G A T A, G A C A repeats. The treble clef musical scores are accompanied by corresponding base sequences, shown as cDNA, and amino acid sequences encodable by one of the three open reading frames. Translation was based upon the mammalian mitochondrial coding system which is thought to be older than the universal coding system[28].

base oligomers, numbers of bases in these oligomeric units not being multiples of three.

Viewed in the above light, it appears that the set of primordial genes for sex determination also has to have arisen very early, this time after the division of eukaryotes from prokaryotes, but in unicellular eukaryotes. As far as metazoans with a pair of ovaries and/or testes are concerned, it seems as though too much attention has been paid to the presence or absence of a genetic mechanism for sex determination in a given species. The fact is that differential regulation of the set of ovary-organizing and testis-organizing genes is practiced by both hermaphroditic species and gonochoristic species alike; within each individual in the case of former and between individuals (males and females) in the case of latter.

The CAM (cell adhesion molecule) which causes aggregation of the slime mold, as well as the N-CAM involved in neuronal organization of the chicken, appear to have evolved from simple Thr-X, Thr-X dipeptidic repeats. Similarly, I believe that the CAM involved in ovarian and testicular organogenesis evolved from simple tetrapeptidic repeats encoded by T A T C and T G T C repeats. I have paid tribute to this ultimately primordial coding sequence by transforming it into a musical score.

REFERENCES

1. Ohno, S: Evolution by Gene Duplication. Heidelberg: Springer-Verlag, 1970.

2. Dixon, RA, Kobilka, BK, Strader, DJ, Benovic, JL, Dohlman, HG, Frielle, T, Bolanowski, MA, Bennett, CD, Rands, E, Diehl, RE, Mumford, RA, Slater, EE, Sigal, IS, Caron, MG, Lefkowitz, RJ, Strader, CD: Cloning of the gene and cDNA for mammalian -adrenergic receptor and homology with rhodopsin. Nature, 321:75-79, 1986.

3. Dayhoff, MO: (editor) Atlas of Protein Sequences and Structure. Silver Springs, Maryland: Nat Biomed Res Found, 1972

4. Wakabayashi, S, Matsubara, H, Webster, DA: Primary sequence of a dimeric bacterial haemoglobin from Vitreoscilla. Nature, 322:481-483, 1986.

5. Ohno, S: Immortal Genes. Trends in Genet, 1:196-200, 1985.

6. Jenni, L, Marti, S, Schweizer, J, Betschart, B, Le Page, RWF, Wells, JM, Tait, A, Paindavoine, P, Pays, E, Steinnert, M: Hybrid formation between African trypanosomes during cyclical transformation. Nature, 322: 173-175, 1986.

7. Ohno, S: Animal Cytogenetics: Protochordata, Cyclostomata and Pisces. Berlin and Stuttgart: Gebruder Borntrager, 1974

8. Shapiro, DY: Sequence of coloration changes during sex reversal in the tropical marine fish Anthias squaminpinnis (Peters). Bull Marine Science 31:383-398, 1981.

9. Pieau, C: Intersexuality in the Animal Kingdom. Reinboth C., ed. Berlin: Springer-Verlag, 1975.

10. Pieau, C, Dorizzi, M: Determination of temperature sensitive stages for sexual differentiation of the gonads in embryos of the turtle, Emys orbicularis. J Morphol, 170:373-382, 1981.

11. Zaborski, P, Dorizzi, M, Pieu, C: Sur l'utilization de serum anti-H-Y de soures pour la determination du sexe genetique chez Emys orbicularis (Testudines, Emydidae). C R Acad Sci (Paris) 28:351-354, 1979.

12. Edelman, GM: Evolution and morphogenesis: the regulator hypothesis. In: Gustafson, JP, Stebbins, GL, Ayala, FJ, eds Genetics, Development and Evolution (17th Stadler Genetics Symposium). New York and London: Plenum Press, 1986.

13. Hemperly, JJ, Murray, BA, Edelman, GM, Cunningham, BA: Sequence of a cDNA clone encoding the polysialic acid-rich and cytoplasmic domains of the neural cell adhesion molecule N-CAM. Proc Natl Acad Sci USA, 83:3037-3041, 1986.

14. Wachtel, SS, Ohno, S, Koo, GC, Boyse, EA: Possible role of H-Y antigen in primary sex determination. Nature (London) 257:235-236, 1975.

15. Ohno, S, Klinger, HP, Atkin, NB: Human oogenesis. Cytogenetics, 1:42-51, 1962.

16. Ohno, S: Major sex determining genes. Berlin-Heidelberg-New York. Springer-Verlag, 1979b

17. McLaren, A: Mammalian Chimaeras, Development and Cell Biology 4. London: Cambridge Univ, 1976

18. Yamamoto, T: Progenies of sex-reversal females mated with sex-reversal males in the medaka, Oryzias latipes. J Exptl Zool, 146:163-180, 1961.

19. Mikamo K. and Witschi E. Masculinization and breeding of the WW Xenopus. Experientia, 20:622-624, 1964.

20. Miller, RA: Spermatogenesis in a sex-reversed female and normal males of the domestic fowl, Gallus domesticus. Anat Rec, 70:155-189, 1938.

21. Noegel, A, Gerisch, G, Stadler, J, Westphal, M: Complete sequence and transcript regulation of a cell adhesion protein from aggregating Dictyostelium cells. Embo J, 5:1473-1476, 1986.

22. Hoyle, F: Ten Faces of the Universe. London:Freeman Press, 1979.

23. Dyson, F: Origins of Life. London-Cambridge, Cambridge University Press, 1985.

24. Miller, SL, Orgel, LE: The Origin of Life on the Earth. New York; Prentice-Hall, 1974.

25. Bridson, PK, Orgel, LE: Catalysis of accurate poly (C) directed synthesis of 3'-5' linked oligoguanylates by Zn^{2+}. J Mol Biol, 144: 567-577, 1980.

26. Eigen, M, Schuster, P: The hypercycle. A principle of natural self-organization. Part A: Emergence of the hypercycle. Naturwiss, 64:541-565, 1977.

27. Hill, D, Blumenthal, T: Does Q replicase synthesize RNA in the absence of template? Nature, 301:350-352, 1983.

28. Ohno, S, Epplen, J: The primitive code and repeats of base oligomers as the primordial protein-encoding sequence. Proc Natl Acad Sci USA, 80:3391-3395, 1983.

29. Singh, L, Purdom, IF, Jones, KW: Conserved sex-chromosome-associated nucleotide sequences in eukaryotes. In: Cold Spring Harbor Sym Quant Biol 45:805-814, 1981.

30. Ohno, S, Ohno, M: The all-pervasive principle of repetitious recurrence governs not only coding sequence construction but also human endeavor in musical composition. Immunogenetics, 24:71-78, 1986.

poor

SEARCHES FOR HUMAN H-Y BY CONVENTIONAL MEANS, AND HOW IT ALL BEGAN

E. J. Eichwald, C. R. Silmser, C. Jorgensen and J. Shelby

Department of Pathology
University of Utah Medical Center
Salt Lake City, Utah 84132

It would appear that H-Y has two faces, not only in regard to its fields of action, but also in regard to its deportment. On the one hand H-Y is present, and makes itself persistently felt, throughout the vertebrate phylum, wherever there are heterogametes[1]; it is all-pervasive, and sees to it that a second gender arises; it sees to it that the genus, be it platyfish, rattlesnake, hippopotamus, or man does not attain a state of panfeminity. On the other hand, when it comes to immune responses, H-Y is fastidious, highly discriminating, and refuses to take action unless certain well-defined preconditions are met. We have learned from Hurme et al[2], from Greene et al[3], and from Liew et al[4] that even a relatively simple event, e.g. delayed type hypersensitivity (DTH), or the rejection of murine skin grafts, will not come to pass unless responders carry the b allele of the MHC locus and, in addition the b allele of the I-B sublocus. In an equally simple event, namely, the elaboration of cytotoxic cells, the I-B sublocus matters less, and what is important is the I-A sublocus. It is surprising, then, that H-Y was first observed, not while it performed its pervasive sex-determining role, but when the demanding conditions for immune responsiveness were incidentally met[5]. I shall return briefly to this aspect below.

However, responses elicited by H-Y are not as fastidious as they might appear on first glance. Once the response has been set in motion, antibody or cytotoxic cells will bind to antigenic matter of more general provenance[6,7]. If such were not the case, any plans for publications like the present volume might not have materialized. We are told in the paper by Greene, cited above, that the evidence for H-Y restrictiveness in mice is merely "suggestive", and that H-Y responses of I-B^{b} mice are merely "significantly faster" than those of other genotypes. Gordon et al[8] also observed that C57BL/6 females, primed with allogeneic male skin, rejected subsequent grafts of syngeneic male skin with increased speed, a result which indicates that even MHC restriction, if present, is only relative. Furthermore, Silvers et al[9] observed a similar effect after priming with rat skin, and Wiberg's observation on the effect of cultured fibroblasts from male human donors points in a similar direction[10].

Given the present state of incomplete information, it is not surprising that occasionally I am visited by obstetricians or neonatologists who tell me about a patient of theirs whose external genitals do not look as they are expected to look. They are dealing with a case of abnormal sex differentiation, and they request that I run an H-Y test on their patient. At such times I tend to quote Gloria Koo's statement that "H-Y is a weak antigen....represented sparsely on cell surfaces... that serologic tests for H-Y are technically difficult, and that it requires considerable skill and experience to master the sperm cytotoxicity assay"[11]. We do not perform the test that they are requesting, and I do not know of any service laboratory that will. Clearly, one cannot transplant skin from a given patient to a syngeneic, female sibling. Likewise, one cannot with advantage transplant skin from such a case to unrelated male and female recipients in order to determine whether the female recipients reject the skin faster than the male recipients; additional, in part MHC-determined antigens will evoke rejection reactions that overshadow any responses elicited by H-Y.

Nevertheless, these requests by my colleagues are not unreasonable, and over the years we have tried to respond. We aimed at the development of a relatively simple, sensitive, documentable test with a relatively short turn-around time. I shall now report on two of our attempts, one of which has been published[12], while the other one has not.

We turned our attention to measurements of delayed type hypersensitivity (DTH) and to the popliteal lymph node (PLN) assay, as previously reported by Greene et al and others. We intended to progress from the syn- to the allogeneic model, and, if results were promising, from the allo- to the xenogeneic. Table 1 confirms that the PLN assay is positive, even in the absence of recipient priming. Priming of the recipient, as shown in Table 2, appears to intensify the PLN response; foot pad swelling was more intense in the experimental than the control animals, but the difference did not reach statistical significance. When we changed from an HvG (host versus graft) to a GvH (graft versus host) model, by injecting primed female lymph node cells into the foot pads of syngeneic males, both PLN and DTH tests were positive.

Results with allogeneic models were disappointing (see Table 3): although foot pads and popliteal lymph nodes tended to be larger after injection of allogeneic male cells into syngeneically primed females, the differences were not impressive; they were inconsistent, and rarely reached a level of statistical significance. Therefore, we did not proceed to xenogeneic models.

Table 1: Popliteal lymph node responses of non-sensitized syngeneic C57/BL mice to H-Y

Time Interval days	Experimental group	Control group	P
6 - 9 (9)*	3.01 $\pm$ 0.67**	2.29 $\pm$ 0.79	= 0.05
12 (5)	2.60 $\pm$ 0.50	1.28 $\pm$ 0.14	$<$ 0.005
15 - 20 (5)	1.88 $\pm$ 0.34	1.18 $\pm$ 0.12	= 0.01

*: no. mice
**: mg. $\pm$ 1 S.D.

Table 2: Popliteal lymph node (PLN) and foot pad (DTH) responses of sensitized syngeneic mice to H-Y

	n	Exp ± 1 S.D.	Contr. ± 1 S.D.	P
PLN	17	3.32* ± 1.01	2.30 ± .57	<0.005
DTH	17	2.76** ± 0.30	2.48 ± 0.24	>0.10

*) mg
**) mm

A second experiment, published earlier this year, utilized rejection of skin grafts as the endpoint. Gordon et al[8] and Silvers et al[9] had already reported that priming of female mice with allogeneic male cells caused accelerated rejection of subsequent male skin grafts, thereby demonstrating homology of H-Y. Silvers et al had even made a similar observation regarding priming with xenogeneic (rat) cells, a result not in agreement with earlier observations of Billingham et al[13]. These results encouraged us to hope that the skin graft model might supply the answer, provided that the endpoints could be sharpened, and the length of the experiments (which in H-Y models tends to extend beyond 2 weeks) could be shortened. With these aims in mind, we developed a semiquantitative histologic assay which was designed to be more sensitive than mere gross observation, and could conceivably even demonstrate immune responses that fall short in their aim of rejection of the test graft. The assay was also designed to eliminate the unavoidable, subjective element of gross observation, and to yield documentable data after a relatively short turn-around time - an important requirement for service laboratories. The histologic assay was centered around the degree of infiltration by host cells into the graft, particularly the perifollicular areas, and the decay of its epithelial elements, particularly the sebaceous glands. Tables 4, 5, and 6 present a summary of the results of numerous

Table 3: PLN and DTH responses of C57BL/6 ("B6") female mice to allogeneic BALB/c ("C") H-Y

Assay	Pretreatment	Challenge	Exp.	Control	P
PLN	--	C male & fem.	2.28±0.66(12)	1.95±0.59(12)	>.10
"	B6 male	"	4.93±1.30(4)	4.50±0.63(4)	>.10
"	C male & C female	B6 male & fem.	2.64±0.32(11)	2.48±0.40(9)	>.10
DTH	B6 male	C male & fem.	2.73±0.15(4)	2.64±0.09(4)	>.10
"	C male & female	B6 male & fem.	2.77±0.16(11)	2.65±0.09(9)	<.05

*: No. of mice
Values for PLN are in mg ± 1 S.D.
Values for DTH are in mm ± 1 S.D.

Table 4: Histologic responses of C57/BL female mice to syngeneic male grafts

Pretreatment	D8	D12	D15
none	(control female grafts)	-	1.67 ± 1.51*(6)**
none	2.34 ± 2.0 (29)	3.28 ± 2.7(54)	5.24 ± 1.64(41)
i.p. sens.	4.95 ± 3.59(22)	8.49 ± 2.40(59)	8.90 ± 1.70(12)
Newborn inj.	--	--	0.75 ± 0.89 (8)

modifications, described earlier in greater detail[12]: modifications pertaining to dosage; time intervals; provenance of priming cells, - whether syn-, allo- or xenogeneic; and whether the responders were adult or newborn. The results are obvious: the assay works well on syngeneic priming; it works fairly well on allogeneic priming; it does not work at all on xenogeneic priming, be it by rat or man.

Clearly, this attempt to develop an assay was also a failure. However, Wiberg has reported success[10] in priming mice with cells from primary cultures of human fibroblasts obtained from known male and female donors. As yet we have not attempted to confirm these remarkable results.

Why did we fail, and why should it be so difficult to demonstrate the presence of allo- and xenogeneic H-Y? Nobody doubts that mice other than C57BL/6 carry H-Y, that species other than the mouse carry it, and that H-Y moieties of assorted provenance cross-react, and possibly are identical. Is it merely a matter of disparity <u>vs</u> consonance at the MHC, or at an MHC sublocus, between donor and recipient, a disparity which would be high in allogeneic, and even higher in xenogeneic combinations? Probably, other or additional factors play a role. Should one not merely consider that the effectiveness of any priming process depends on the dose of antigen, and the time period during which a potential responder is exposed to it? A strongly antigenic cell will evoke a strong immune response, even if this strong response succeeds in eliminating the antigenic cell after a brief period of exposure. A weakly

Table 5: Histologic responses of C57/BL female mice to H-Y after allogeneic sensitization

Sensitizing cells	D8	D15
BALB/c male	7.27 ± 1.49(11)	9.20 ± 1.97(15)
BALB/c female	2.50 ± 1.52(6)	6.17 ± 1.94(6)
C3H/SW male	6.00 ± 2.39(8)	7.60 ± 1.14(5)
C3H/SW female	3.18 ± 2.60(11)	6.50 ± 0.58(4)
C3H male into NB	--	2.43 ± 2.24(16)

Table 6: Histologic responses of C57/BL female mice to H-Y after xenogeneic sensitization

Sensitizing cells	D8	D12	D15
LEWIS male	1.00±0.82(4)	3.86±3.06(14)	4.33±1.21(6)
LEWIS female	3.83±4.07(6)	3.00±2.06(9)	5.50±1.38(6)
HUMAN WBC, male	3.00±1.78(18)	3.42±1.98(12)	--
HUMAN WBC, female	5.00±1.10(6)	3.20±2.77(5)	--
HUMAN WBC, male into NB	--	--	6.50±1.91(4)

antigenic cell will evoke a weak response, although, because of its weakness, the sojourn of this weakly antigenic cell will be relatively long. Cells which are both strongly and weakly antigenic, e.g. male allo- or xenogeneic cells, will be eliminated earlier the stronger they are. Consequently, weak antigens, straddling allo- or xenogeneic antigenic cells, will be rendered ineffective, being eliminated at an early date.

I have previously described some of the details of the discovery of H-Y[14]. I merely wish to re-emphasize here the role of luck in laboratory work. If, at that time, we had not used mice of the readily available C57BL strain (in order to develop and improve our skin grafting technique), someone else, at some other place, at some other time, would have noted the H-Y effect, soon thereafter. But why, in addition to ready availability did we use C57BL, instead of CBA, or A, or BALB/c mice which also were readily available? We used C57BL mice because we had learned that they did not tend to develop acquired immunologic tolerance via intrauterine or newborn intravenous injection of antigenic cells. We wanted to know why they did not. We should also recognize here another, very significant stroke of good luck which occurred in the laboratories of Billingham, Brent and Medawar. If they had not injected CBA cells into A mice, but had used almost any other genetic combination which involved disparity at the H-2 locus, tolerance might not have been observed, and we might still live in the dark ages when acquired specific tolerance was merely a matter of speculation.

REFERENCES

1. Wachtel, SS: A transplanter's look at gonadal induction. Transpl, 311-3, 1981.

2. Hurme, M, Hetherington, CM, Chandler, PR, Simpson, E: Cytotoxic T-cell responses to H-Y: mapping of the Ir genes. J Exp Med, 147:758-767, 1978.

3. Greene, MI, Benacerraf, B, Dorf, ME: The characterizations of the delayed-type hypersensitivity reaction to H-Y antigen. Immunogenet, 11:267-273, 1980.

4. Liew, FH, Simpson, E: Delayed type hypersensitivity responses to H-Y: characterizations and mapping of Ir genes. Immunogenet, 11: 257-268, 1980.

5. Eichwald, EJ, Silmser, CR: Communication. Transpl Bull, 2:148-149, 1955.

6. Goldberg, EH, Boyse, EA, Bennett, D, Scheid, M, Carswell, EA: Serological demonstration of H-Y (male) antigen on mouse sperm. Nature, 232:478-480, 1971.

7. Koo, G, Varano, A: Inhibition of cell-mediated cytolysis by monoclonal H-Y specific antibody. Immunogenet, 14:183-186, 1981.

8. Gordon, RD, Mathieson, BJ, Samuelson, LE, Boyse, EA, Simpson, E: The effect of allogeneic presensitization of H-Y graft survival and in vitro cell-mediated responses to H-Y antigen. J Exp Med, 144:810-820, 1976.

9. Silvers, WK, Yang, S-L: Male specific antigen homology in mice and rats. Science, 181:570-572, 1973.

10. Wiberg, UH: H-Y transplantation antigen in human XO females. Human Genet, 69:15-18, 1985.

11. Koo, G: Serology of H-Y antigen. Human Genet, 58:18-20, 1981.

12. Worley, R, Jorgensen, C, Eichwald, EJ: Limitations of a histologic assay for syn-, allo-, and xenogeneic H-Y antigen. J Reprod Immunol, 8:121-128, 1986.

13. Billingham, RE, Hodge, BA, Silvers, WK: An estimate of the number of histocompatibility loci in the rat. Proc Nat Acad Sci USA, 48:138-147, 1962.

14. Eichwald, EJ, Silmser, CR: H-Y and serendipity. Human Genet, 58:6-8, 1981.

SEROLOGICAL ANALYSIS OF H-Y ANTIGEN

Gloria C. Koo* and Claude Nagamine#
*Merck, Sharp & Dohme Research Laboratories, Rahway, NJ 07065
#Memorial Sloan-Kettering Cancer Center, New York, NY 10021

INTRODUCTION

H-Y antigen has attracted the attention of two major classes of investigators: the cellular immunogeneticists, who originally defined the H-Y antigen system; and the reproductive biologists, who have studied H-Y as the product of a gene on the Y chromosome and investigated its possible role in sex differentiation. Spanning the boundary between the two disciplines is the serologist of H-Y. Indeed, the serological detection of H-Y and the subsequent studies of cross-reactivity of H-Y from human and other species initiated the explosion of literature on H-Y as the determinant for heterogametic sex. Such studies led to the elucidation of many aspects of sexual differentiation and focussed attention on the Y chromosome in numerous abnormal and unusual cases of sex differentiation in man and other species[1]. It has become clear that, although the male phenotype is always associated with H-Y, the converse is not always true: XX, XO, and XY females can also express H-Y, depending on the nature of the abnormalities in their genetic constitution. As sexual differentiation is a complex process, it is not surprising that H-Y is one of many gene products involved in primary sex determination. Since the experimental basis of many hypotheses about the role of H-Y antigen is dependent on serological determination of H-Y, serological assays of H-Y and their problems merit considerable attention.

SEROLOGICAL DETECTION OF H-Y

H-Y was originally defined by transplantion techniques, when tissue (e.g. skin) from males was rejected by females of the same inbred strain[2]. Serological detection of H-Y was first performed by Goldberg et al, using antiserum raised in females which rejected male skin grafts and mouse sperm as targets for a complement dependent cytotoxicity assay[3]. The various serological assays used to detect H-Y corresponded with the state of the art for other antigenic systems at the particular time when they were designed. These assays have been reviewed previously[4].

H-Y is a weak antigen and the antigenic determinant appears to be a carbohydrate moiety[5]. In our experience, the most accurate and sensitive tests for H-Y make use of homogeneous populations of cells. For the mouse system, epididymal sperm, Concanavalin A-stimulated blast cells (mainly T cells) and fibroblasts are a good source of target cells. To test for antigenicity of cells from other species, we have indirectly absorbed the H-Y antiserum with the cells in question. The removal of H-Y reactivity indicates that H-Y antigen is present on the cells[4]. In the human system, this indirect method is widely used. Useful targets include mouse sperm and human Raji cells, which are male lymphoma cells that apparently express sufficient H-Y for detection. In our laboratory, we have developed radiobinding assays, using radiolabeled primary or secondary reagents to detect H-Y antigen on human or mouse cells[6]. We have performed direct radiobinding tests on human fibroblasts and found them to be quite suitable for testing. Most recently an ELISA (enzyme-linked-immunosorbant-assay) was developed, using culture supernatant which contains the presumptive, soluble H-Y as the antigen[7]. This assay appears very sensitive and the difference between male and female samples is very impressive.

Serology of any weak antigen is difficult, and assays for the presence of H-Y are no exception. Source of antibody, preparation of target cells and various secondary reagents are of prime importance. A serious investigator has to take all three parameters into consideration.

MALE SPECIFICITY OF H-Y ANTISERUM

The male-specific reactivity of H-Y antiserum has often been challenged[8], because of the variety of serological tests used to detect H-Y. The best evidence that H-Y antiserum is detecting a male-specific antigen appears to come from serological tests used to sex embryos or fetuses. Kroc & Goldberg were the first to detect H-Y on 8-cell mouse embryos[9] and Epstein et al. were able to identify the H-Y (+) embryos correctly as male[10]. These findings were subsequently repeated by White et al[11]. We have used monoclonal H-Y antibodies in tests with human amniotic fluids, to correlate their reactivities with the sex of the fetus from whom the fluids were derived. Out of 100 coded samples, we correctly identified 90% of the male fetuses & 100% of the female fetuses[12]. These data unequivocally demonstrate the male-specific reactivity of H-Y antiserum.

The reactivity of H-Y antibody with XO cells still presents a problem. Reports in the literature are rather controversial: XO mice have been found, in separate studies, to be positive and negative in assays with H-Y antiserum[13,14], and by H-Y cell-mediated cytotoxicity[14,15]. The same conflicting results have been found for human XO individuals[4,16,17], although the human situation could be complicated by the presence of Y chromosomal material which is undetectable by conventional karyotyping. One explanation for the reactivity of XO cells was offered by Adinolfi et al[18], who proposed the existence of a precursor molecule, H-Yp, that is encoded by autosomal genes and converted to the antigenic H-Y, by Y-chromosomal gene, by glycosylation. They postulated that the X chromosome also encodes a product that can convert H-Yp by glycosylation to another molecule related to H-Y, namely H-Yr. Accordingly, XX females convert more H-Yp to H-Yr than the XO females and the remaining H-Yp cross-reacts with H-Y antigen. Although the hypothesis is attractive, more biochemical data are required to elucidate the situation in XO cells.

Nevertheless, the male-specific reactivity of anti-H-Y antiserum has found a practical application in the sexing of embryos and sperm for research in animal sciences. Encouraging results are certainly evident in the bovine system[19,20].

THE H-Y ANTIGEN DETECTED BY SEROLOGY AND BY CELLULAR ASSAYS IS PROBABLY THE SAME.

When inbred female mice are sensitized by male mice, both humoral and cellular immune responses are invoked. The humoral responses are detected by antibody and the various serological assays described above. The cellular responses are detected by transplantation[2], cell-mediated cytotoxicity[21] and popliteal lymph node enlargement[22]. All these processes are complex, involving the antigen-presenting cells (eg. macrophages), T cells and B cells of the immune system, which are under control of the major histocompatibility complex[23]. We believe that the same H-Y antigen is recognized by each of these different immunological assays. T cells recognise a processed antigen in the context of MHC antigen, but the antibody recognises antigenic sites on the whole molecule[23]. In the past, these questions were approached by performing blocking assays _in vivo_. For example, antibodies were injected into mice which were receiving transplants, or antibodies were used to block a cellular assay _in vitro_. Antibodies to H-Y have been shown to modulate rejection of skin grafts, either by enhancing or delaying a graft[24]. We have been able to block lysis of T cells specific for H-Y and the popliteal lymph node response with monoclonal antibodies[25,26]. These data suggest that the same epitopes are recognized by both humoral and cellular responses. Definitive answers on the nature of these epitopes will become available when H-Y antigen is biochemically characterized and the H-Y gene is cloned.

FACTORS AFFECTING IMMUNOGENICITY OF H-Y

Several factors could affect the antigenicity of H-Y, as follows.
(a) The major histocompatibility complex (MHC): MHC affects the immune responses of many antigens, being a part of the antigen-recognition component of the cell (see ref. 23). Since H-Y is probably processed in the same way as any other antigen, the Class II MHC determinant involved in antigen-presentation must be important in initiation of anti-H-Y immune responses. In fact, in both skin graft transplantation and cytotoxic T cell assays, MHC controls both the immunogenicity and the immune response of H-Y[21,27].
(b) Beta-2-microglobulin (B2-m): B2-m may influence the antigenicity of H-Y. This observation was made by several laboratories in experiments with the human male lymphoma cell line, Daudi, which lacks B2-m and secretes H-Y into the culture medium[6,28]. However, no biochemical linkage has yet been established, although the findings suggest that levels of H-Y could be quantitatively influenced by B2-m.
(c) Secretion of H-Y: H-Y is secreted into the medium by Daudi and other cell types, such as Serotoli cells and testicular cells[6]. These results indicate that H-Y may be only loosely inserted into the cell membrane. This physiological property of H-Y may create problems in handling cells for determination of H-Y.
(d) Hormonal effects: In both the skin transplantation and serological assays for H-Y, testosterone appears to influence the antigenicity[29] or "mobility" of H-Y on the cell surface[30]. Recently hormones have

also been shown to affect the expression of H-Y on avian cells[31,32].

All these factors add to the complexity of an already weak system and may all influence serological determinations of H-Y antigen.

CONCLUDING REMARKS

Our brief review of the vast literature on the serology of H-Y is designed solely to emphasize the following issues:

(1) Serological determination of H-Y is male-specific and achievable. Reagents, such as monoclonal antibodies and target cells, are of prime importance.

(2) H-Y antigen detected by serology and cellular assays is most probably the same molecule.

(3) Biochemical identification of H-Y antigen and the subsequent cloning of the H-Y gene will ultimately resolve the controversy that surrounds the serology of the H-Y antigen.

ACKNOWLEDGMENT

We thank Ms. Dawn Smith for excellent secretarial assistance.

REFERENCES

1. Wachtel, SS: H-Y Antigen and the Biology of Sex Determination, Grune and Stratton N.Y. 1982.

2. Eichwald, EJ, Silmser, CR: Untitled communication. Transplant Bull, 2:148, 1955.

3. Goldberg, EH, Boyse, EA, Bennett, D, Scheid, M, Carswell, EA: Serological demonstration of H-Y (male) antigen on mouse sperm. Nature (London), 232:478, 1971.

4. Koo, GC: Serology of H-Y antigen. Human Genetics, 58:18, 1981.

5. Shapiro, M, Erickson, RP: Evidence that the serological determinant of H-Y antigen is carbohydrate. Nature, 290:503, 1981.

6. Nagamine, C, Reidy, J, Koo, GC: A radiobinding assay for human H-Y antigen using monoclonal antibodies. Transplantation, 37:14, 1984.

7. Brunner, M, Moreira-Filho, CA, Wachtel, G, Wachtel, S: On the secretion of H-Y antigen. Cell, 37:615, 1984.

8. Crichton, DN, Steel, CM: Serologically detectable H-Y (male) antigen: Mr. or Myth? Immunol Today, 6:202, 1985.

9. Kroc, CJ, Goldberg, EH: Detection of H-Y (male) antigen on 8-cell mouse embryos. Science, 193:1134, 1976.

10. Epstein, CJ, Smith, S, Travis, B: Expression of H-Y antigen on preimplantation mouse embryos. Tissue Antigens, 15:63, 1980.

11. White, KL, Lindner, GM, Anderson, GB, BonDurant, RH: Cytolytic and fluorescent detection of H-Y antigen on preimplantation mouse embryos. Theriogenology, 19:701, 1983.

12. Koo, GC, Reidy, JA, Hammerling, U, Cederqvist, LL: Fetal H-Y typing using human amniotic fluid. J Reprod Immunol, 3:59, 1983.

13. Engel, W, Klemme, B, Ebrecht, A: Serological evidence for H-Y antigen in XO-female mice. Hum Genet, 57:68-70, 1981.

14. Koo, GC, Reidy, JA, Nagamine, CM: H-Y antigen in XO mice. Immunogenetics, 18:37, 1983.

15. Simpson, E, McLaren, A, Chandler, P: Evidence for two male antigens in mice. Immunogenetics, 15:609, 1982.

16. Wolf, U, Fraccaro, M, Mayerova, A, Hecht, T, Zuffardi, O, Hameister, H: Turner syndrome patients are H-Y positive. Hum Genet, 54:315, 1980.

17. Wiberg, UH: H-Y transplantation antigen in human XO females. Hum Genet, 69:15, 1985.

18. Adinolfi, M, Polani, P, Zenthon, J: Genetic control of H-Y synthesis: A hypothesis. Hum Genet, 61:1, 1982.

19. White, KL, Bradbury, MW, Anderson, GB, BonDurant, RH: Immunofluorescent detection of a male-specific factor on preimplantation bovine embryos. Theriogenology, 21:275, 1984.

20. Ali, JJ, Eldridge, FE, Anderson, GA, Koo, GC, Schanbacher, BD: Separation of bovine X- from Y-chromosome bearing spermatozoa using monoclonal H-Y antibody. 1986-submitted for publication.

21. Simpson, E, Gordon, RD: Responsiveness to H-Y antigen. Ir gene complementation and target cell specificity. Immunol Rev, 35:59, 1977.

22. Johnson, LL, Bailey, DW, Mobraaten, LE: Genetics of histocompatibility in mice, III. Characteristics of the popliteal lymph-node response to minor (non-H-2) H antigens. Immunogenetics, 13:327, 1981.

23. Schwartz, RH: The role of gene products of the major histocompatibility complex in T cell activation and cellular interactions. In: Fundamental Immunology ed. W. E. Paul, Raven Press, N.Y. 379, 1984.

24. Wachtel, GM, Wachtel, SS, Nakamura, D, Moreira-Filho, CA, Brunner, M, Koo, GC: H-Y antibodies recognize the H-Y tranplantation antigen. Tranplantation, 37:8, 1984.

25. Koo, GC, Varano, A: Inhibition of H-Y cell-mediated cytolysis by monoclonal H-Y-Specific antibody. Immunogenetics, 14:183, 1981.

26. Jeong, G, Reidy, JR, Koo, GC: Inhibition of H-Y specific popliteal lymphocyte response with monoclonal antibodies. Transplant, 41:126, 1986.

27. Wachtel, SS., Gasser, DL, Silvers, WK: An association between H-2 and the expressivity of H-Y in murine skin grafts. Transplant Proc, 5:295, 1973.

28. Beutler, B, Nagai, Y, Ohno, S, Klein, G, Shapiro, JM: The HLA - dependent expression of testis-organizing H-Y antigen by human male cells. Cell, 13:509, 1978.

29. Polackova, M, Vojtiskova, M: Inhibitory effect of early orchidectomy on the expression of the male antigen in mice. Folia Biol (Praha), 14:93, 1968.

30. Flaherty, L, Zimmerman, D, Wachtel, SS: H-Y antigen: Cell surface mapping and testosterone - induced supramolecular repatterning. J Exp Med, 150:1020, 1979.

31. Zaborski, P, Guichard, A, Scheib, D: Transient expression of H-Y antigen in quail ovotestis following early diethylstilbestrol (DES) treatment. Biol Cell, 41:113, 1981.

32. Koo, GC, Allen, L, Long, RA, Serio-Dunn, R, Goggin, B, Weppelman, RM: Effect of Tamoxifen on H-Y antigen expression and gonadal development in chicken embryos. Differentiation, 29:140, 1985.

IMMUNOGENETIC ANALYSIS OF THE H-Y ANTIGEN

Ellen H. Goldberg and Brian Reilly

Department of Microbiology
University of New Mexico
School of Medicine
Albuquerque, New Mexico 87131

HISTORICAL PERSPECTIVE

INTRODUCTION

The H-Y histocompatibility antigen was originally described as a cell surface component, responsible for the rejection of male skin grafts by otherwise histocompatible females of the same inbred strain[1,2]. It seemed reasonable at the time to call the gene responsible for the expression of this male-specific antigen H-Y, because the incompatibility between the male and female had to be somehow linked to the Y chromosome[3,4].

Until 1971 there had been no conclusive serological demonstration of the antigen, which limited further analysis of the H-Y system. At that time, we modified the complement-dependent, cytotoxic assay for the detection of major histocompatibility antigens on the surface of mouse spermatozoa, which was then adapted for detection of other antigens on the surface of sperm. Thus, using serum collected from C57BL/6(B6) females which had rejected male skin, we demonstrated the presence of H-Y on mouse sperm[6]. It became apparent that the H-Y antigenic system was unique in that the antigen, recognized serologically, and also by transplantation, appeared to be largely conserved throughout the evolution of vertebrates[7,8]. For this reason, and also because it is expressed on the 8-cell, male mouse embryo[9,10,11], it was suspected that H-Y was a single molecular species involved in primary sex determination of the male[12]. However, there is now increasing evidence suggesting multiplicity or heterogeneity of H-Y molecules[13,14,15], so that detailed characterization of this molecular species in structural terms is now necessary. Furthermore, the role, originally assigned to H-Y, of the inducer of testes needs to be reevaluated in light of some recent data from experiments with sex reversed mice (i.e. XX males and XY females). The characterization of the molecule is therefore relevant to an assessment of its function and to further understanding of this highly conserved piece of genetic information. This report presents a historical perspective of the H-Y antigenic system and includes the current status of biochemical studies which focus on the structural characterization of H-Y.

Immunogenetic Analysis

The H-Y antigenic system is a particularly useful model for studies of the genetic control of immune responsiveness, and the cellular and humoral responsiveness of tissues transplanted across minor histocompatibility antigenic barriers. It was demonstrated that male graft rejection is under the control of a gene or genes within the I region of the major histocompatibility complex[16,17]. Thus, B6(H-2^b) females always reject male skin; in the DBA(H-2^d) strain, male-to-female grafts are rejected by only some recipients; and C3H(H-2^k) females usually accept male skin grafts indefinitely. The first indication that the failure of female mice of certain strains to reject male skin was not attributable to complete absence of an H-Y immune response was the finding that females of "non-rejector" strains, such as C3H, make H-Y antibody even though they fail to reject male skin[18]. Further, second male grafts were often rejected while the first graft was retained. In addition, C3H females often become rejectors when thymectomized at 8 weeks of age[19], splenectomized[20], or past reproductive age[19]. To provide a situation in vitro that parallels male skin graft rejection, we modified the cell-mediated cytotoxicity (CMC) assay for the lysis of male target cells by H-Y specific, cytotoxic T lymphocytes (CTL)[21]. In this case, spleen cells from B6 female mice, grafted with male skin, were the source of H-Y specific CTLs which were reacted with ^{51}Cr-labeled male and female (as a specificity control) lymph node cells as targets. The results clearly indicated a male-specific lysis of the cells. This assay was further modified by Gordon and his colleagues[22] who extended these studies to show that susceptibility of the male target cell to lysis was restricted by the major histocompatibility complex (MHC)[23]. When the CMC assay was applied to detection of H-Y on male cells, using CTLs from the "non-rejector" mice (i.e. H-2^k, CBA, and C3H strains), an H-Y specific response was only elicited if the route of immunization was the footpad (rather than by intraperitoneal injection)[24], or if the females were thymectomized 6 weeks prior to the intraperitoneal primary immunization with male cells[25].

At the time when these studies were performed, the H-Y antigen which had been identified by transplantation techniques, and the H-Y antigen which had been recognized serologically, were presumed to be identical. The H-Y specific antiserum originally used in the serological assays had been elicited by grafting B6 female mice with male skin. Furthermore, using the H-Y specific, MHC-restricted, CMC assay, Koo and Varano[26] were able to inhibit lysis of male target cells with H-Y specific antibody. The assumption that the antigens were identical was questioned, however, as a result of a study in which a single male XO mutant mouse, described by Melvold, was shown by transplantation studies to be H-Y negative (each of 13 female mice permanently accepted skin donated by this mutant) and yet the XO mouse was found to be H-Y positive when tested serologically[13]. These findings indicated that the two assays detect different forms of the H-Y antigen. It was proposed that the antigen recognized by transplantation and cell-mediated assays be designated H-Y and that the antigen recognized by serological techniques be called SDM (serologically defined male antigen[14]). In addition, the studies just described, as well as more recent findings which revealed the serological expression of H-Y by XO female mice[27,28], demonstrated that the structural gene for the serologically defined H-Y antigen is not found on the Y chromosome. However, these conclusions have been challenged by Ohno[29], who argues that murine H-Y antiserum is generally of very low titer, used generally at a 1/8 or 1/16 dilution. Because of this low titer, it was suggested that there are antibodies in the serum which react non-specifically with the target tissue. Ohno has demonstrated that carefully selected antisera, raised in B6 females immunized with male tissue, are clearly H-Y specific and do not react with cells of the XO genotype.

Whether or not H-Y and SDM are the same or different molecules should be resolved conclusively by the use of monoclonal antibodies, biochemical analysis and ultimately by cloning of the gene(s). Resolution of this problem will have considerable impact on discussions of the role played by H-Y in male differentiation.

CURRENT STATUS

Role played by H-Y antigen in male differentiation

The presence of a Y chromosome is normally associated with testicular differentiation in mammals, since XY, XXY and XYY individuals almost always develop as males, whereas XX and XO individuals differentiate as females. For this reason, it has been hypothesized that a gene (or genes) on the Y chromosome is the primary sex determinant involved in testis differentiation. Wachtel et al originally proposed that the gene responsible for the induction of testicular differentiation also coded for the H-Y antigen, as detected by serological assays[30]. Support for their hypothesis was found in results of experiments in which we demonstrated that the H-Y antigen was found on the male mouse embryo at times as early as the 8-cell-stage of development. Using the complement-dependent cytotoxicity assay, we showed that 50% of mouse embryos treated with H-Y antibody and complement were lysed[9]. We assumed that the lysed embryos were indeed male, but, as pointed out by Erickson[31], because of the insensitivity of the assay, our assumption may not have been statistically valid. For this reason, it was critical to show that the embryos lysed by treatment with H-Y antibody were, in fact, male, and this critical fact was demonstrated by 2 independent laboratories as well as our own. Epstein and his colleagues[10] showed by chromosomal analysis that the majority of the unaffected embryos treated with H-Y antibody and complement were XX females. Our laboratory[11] and White et al[32] showed that over 80% of offspring, which survived treatment with H-Y antibody and complement at the blastocyst stage and transfer to the uteri of pseudopregnant mothers, were female. Further support for the hypothesis was presented by Ohno et al[33], and Zenzes et al[34], who demonstrated that dispersed testicular cells from rodents, when reaggregated, formed cord-like structures similar to seminiferous tubules. However, when the testicular cells were cultured in the presence of H-Y antibody, development of tubule-like structures did not occur. Further experiments, using dissociated ovarian cells from newborn rats, showed that when these cells reassociated in the presence of a source of H-Y antigen (in this case, supernatant fluid from rat testicular cells) they formed tubular structures typical of testicular organization[35]. This finding was confirmed by Muller and Urban[36] and was extended to demonstrate that the effect of incubation with H-Y supernatant fluid was abrogated in the presence of H-Y specific antiserum.

The role of the Y chromosome was challenged by the discovery of exceptional situations in which the presence of a Y chromosome does not necessarily direct differentiation of the indifferent gonad into a testis. As summarized by Eicher[37], there are probably at least 3 autosomal loci involved in normal testicular development. One locus of note is the Tas (T-associated sex reversal) gene. A mutation at this locus on chromosome 17 was shown by Washburn and Eicher[38] to cause the gonads of XY individuals to differentiate as ovaries or ovotestes. Whether or not these animals are H-Y positive in serological studies has not yet been determined, although they have been shown to be H-Y positive in transplantation studies. Furthermore, Eicher and her colleagues[40] produced XY female mice, none of which developed normal testes, by transferring the Y chromosome from <u>Mus poschiavinus</u> to the B6 genomic background. The mice either developed as females, with two ovaries, or as hermaphrodites. The females are rarely if ever fertile, and are H-Y positive in transplantation experiments and CMC assays. The hermaphrodites, which appear to have normal male genitalia, are fertile. As pointed out by the authors, as well as by Ohno[29], the origin of the Y chromosome in these mice is the feral mouse species, <u>Mus poschiavinus</u>, which was placed in the B6 genomic background. The <u>H-Y positive, XY</u> female phenotype appears to be strain specific, since the Y^{pos} chromosome on a BALB/c background continues to be male-determining. Therefore, the Y-linked testis determining gene (designated Tdy) on the Ypos chromosome may, in this case, be interacting abnormally with the X-linked or autosomal genes of the B6 mouse which are required for normal testicular differentiation. Thus, it is Ohno's contention that abnormal association between the B6 class I MHC antigen and H-Y antigen of <u>Mus poschiavinus</u> renders H-Y nonfunctional.

Cattanach[41] described a group of mice which carry the sex-determining locus of the Y chromosome as a segment, designated Sxr, which was transferred, by crossing-over during male meiosis, from the Y to the X chromatid[42]. These mice are genotypically XX, but are indisputably male. XYSxr carrier males are now known to carry an extra copy of a region of the Y chromosome, distal to the XY pairing region. Since XXSxr mice develop as males, this region may contain the testis-determining factor, designated Tdy (see above). When the XXSxr male mice were tested for the presence of H-Y by serological[43] and transplantation techniques[44], they were found to be positive, a result which suggests that the Tdy and H-Y genes are identical.

McLaren and her co-workers[45] have demonstrated the presence of the H-Y transplantation antigen in XXSxr females, in which the Sxr-carrying chromosome is partnered by the T(X;16)16H X-autosome translocation, which is differentially active. In this case, T16H/X Sxr mice may develop as females, males or hermaphrodites, suggesting mosaicism and varying degrees of X-inactivation. However, T16H/XSxr females are fertile and the presence of H-Y does not appear to affect their reproductive ability[45].

One female mouse resulting from McLaren's XXSxr stock was found to be a variant, in that she was H-Y negative and yet, when mated to a normal XY male, gave birth to XXSxr males which were H-Y negative, as determined by cell-mediated cytotoxic assays <u>in vitro</u> and transplantation experiments <u>in vivo</u>[47]. This result led to the suggestion that there is a variant form of Sxr which retains the testis determining region, but has lost the genes responsible for expression of the H-Y antigen. This variant was designated Sxr'[45]. In fact, further tests with XOSxr' mice revealed the absence of spermatogenesis and lead to the attractive hypothesis that the genes for spermatogenesis and for expression of the H-Y antigen are the same[46].

Although the findings described here indicate that H-Y, at least as detected by transplantation and CMC assays, is not the primary sex determinant, it remains to be determined whether or not H-Y (SDM), as detected by serological techniques, is involved in testis differentiation.

To resolve some of the unanswered questions (such as the relationship between the different forms of male-specific antigens and their role, if any, in male gonadal differentiation) and to further our understanding of this highly conserved piece of genetic information, the molecule(s) needs to be isolated and characterized.

Biochemical characterization

One of the first attempts[48] to characterize the H-Y antigen biochemically exploited the fact that cultured Daudi cells, originally derived from the human Burkitt lymphoma, produce and secrete H-Y antigen into the culture medium. After Daudi cells were cultured in the presence of tritiated proteins for 16 hours, the medium in which the cells were grown, referred to below as Daudi cell supernant, was concentrated and chromatographed on a Sephadex G-200 column to purify the labeled proteins. The elution profile yielded four major "molecular weight fractions", which were tested individually for the presence of H-Y antigen in a receptor binding assay[48,49]. After reaction of an aliquot from each fraction with bovine fetal ovarian cells (BFO), which had been previously shown to bind H-Y antigen, the amount of bound radioactivity was determined. Most of the proteins which absorbed to BFO were found in the fraction which contained large aggregates of hydrophobic molecules. To permit further characterization of the molecule(s) which bound to the H-Y receptor on the BFO, this fraction was divided into two equal aliquots, one of which was allowed to absorb to BFO while the other was used as a control. The aliquots were then subjected to sodium dodecyl sulfate (SDS) polyacrylamide gel electrophoresis under stringent reducing conditions. Upon comparison of the 2 electrophoretic profiles, it was clear that a protein of molecular weight 18,000 daltons had been selectively removed from the fraction which had been allowed to react with BFO. It was concluded that this 18,000 mol. wt. molecule was the H-Y antigen, since only a single species of protein had attached to the ovarian target cells and, more importantly, ovarian cells to which the 18,000 mol. wt. protein was bound were able to absorb H-Y antibody.

These experiments were extended by Wachtel and his colleagues[50]. They immunoprecipitated a protein of molecular weight 15,000 - 18,000 daltons from a Daudi cell supernatant, using an H-Y specific antibody. However, as they pointed out, the peak which represented the protein of interest sometimes merged with background peaks of proteins of higher molecular weight, so that the results were somewhat difficult to interpret. This problem, however, was overcome when protein A-coated Sepharose beads were used to immunoprecipitate the H-Y immune complex, in place of goat anti-mouse immunoglobulin, thereby eliminating excess immunoglobulin in the preparation[50].

Recently, Bradley and Heslop[51], also using Daudi cell supernatant, immunoprecipitated H-Y antigen with specific antibody and identified 3 predominant proteins with estimated molecular weights of 20,000, 50,000 and greater than 200,000 daltons.

Thus, as demonstrated by 3 independent laboratories, there is a protein of molecular weight 15,000 - 20,000 daltons secreted by Daudi cells, which appears to react specifically with H-Y antibody.

It was shown by Shapiro and Erickson[52] that the antigenic specificity of the H-Y antigen resides in a carbohydrate moiety. This finding has since been confirmed and extended by our use of a monoclonal antibody against H-Y to demonstrate that the serological determinant is a glycoconjugate which contains terminal, nonreducing and internal galactosyl residues[53]. Since Daudi cells derive from a human male lymphoma, it is critical to demonstrate that the serologically recognized moiety is truly the H-Y antigen and not a closely related, tumor-specific antigen which is carbohydrate in nature. Although it is unclear at this time whether H-Y is a single or multiple molecular species (as discussed earlier), it is appropriate to note here that early testing for H-Y by Eichwald et al[4] demonstrated its presence on cells of all normal, male tissues tested, while it was conspicuously absent from most tumors of male background.

Recently, we developed techniques for the isolation and characterization of the H-Y antigen without the use of H-Y antibody for immunoprecipitation. Basing our efforts on existing evidence that H-Y antigen can be shed or secreted, we began testing a variety of sources for soluble H-Y. We had previously observed that suspensions of male splenic lymphocytes, kept at room temperature for 4 hours, no longer reacted with H-Y antibody in the cellular radioimmunobinding assay.

Examination of the supernatant medium, in which these cells had been grown, revealed that: (a) a molecule or molecules in the male lymphocyte supernatant (not female) inhibited the lysis of spermatozoa in the presence of H-Y antibody and complement; (b) supernatant medium from male lymphocytes was capable of eliciting H-Y specific CTL in B6 females; and (c) prior inoculation of supernatant from growth of male lymphocytes into B6 females accelerated the rates of rejection of subsequent male skin grafts. As shown in Figure 1, when H-Y antiserum is diluted in supernatant fluid from the equivalent of 50×10^6 male lymphocytes per ml, there was significant inhibition of lysis of spermatozoa. As a control, H-Y antiserum was diluted in supernatant from an equivalent number of female lymphocytes. We then prepared analogous male and female supernatant fluids from spleens of B6 mice at a concentration of 50×10^6 cells per ml of M199 medium. The supernatants were centrifuged at 100,000 x g to remove cellular debris and B6 females were innoculated with 1 ml of supernatant. Two to seven weeks later (at weekly intervals) the spleens of these females were removed and tested for the presence of H-Y specific CTL *in vitro*. As shown in Table 1, supernatant fluid prepared in this way from male spleen cells contains factors which induce the generation of H-Y specific CTL in B6 females, as demonstrated by a 20%-32% specific lysis of male target cells, at an 8:1 effector:target ratio. Percent specific lysis was determined by the formula: $\%\ \text{specific lysis} = 100 \times \frac{A-B}{C-D}$ where A=^{51}Cr released from target cells incubated with sensitized effector cells, B=^{51}Cr release from target cells incubated with unsensitized effector cells, C=^{51}Cr released from cells treated with Zap-Globin II for maximum lysis and D=machine background.

1 ml of the prepared supernatant was then used to inoculate intraperitoneally (ip) virgin female B6 mice. Four weeks later, the mice were grafted with male skin and the survival times of the graft recorded. As shown in Table 2, experimental B6 females which had received male supernatant fluid rejected the male graft in 24.4 days, whereas control B6 females which had received female supernatant prior to grafting with male skin rejected the graft in 35.2 days. These results show that the spleen cells from male mice release factor(s) which prime females to the male antigen involved in transplantation rejection.

These findings indicate that a factor or factors shed from male lymphocytes can inhibit the serological detection of H-Y and also induce cellular, H-Y specific CTL which is demonstrable *in vivo* and *in vitro*.

In addition to material derived from lymphocytes, we also tested supernatant fluid from cultured male lung fibroblast cells. Lung fibroblasts from B6 male and female mice were grown to confluence under standard culture conditions. The supernatant medium was then removed and examined for H-Y activity by inhibition of the Protein A-sheep red blood cell rosette assay (PrA-SRBC)[54]. As shown in Figure 2, inhibition of rosettes was obtained with male, but not with female supernatant fluid. Because of the ready availability of large

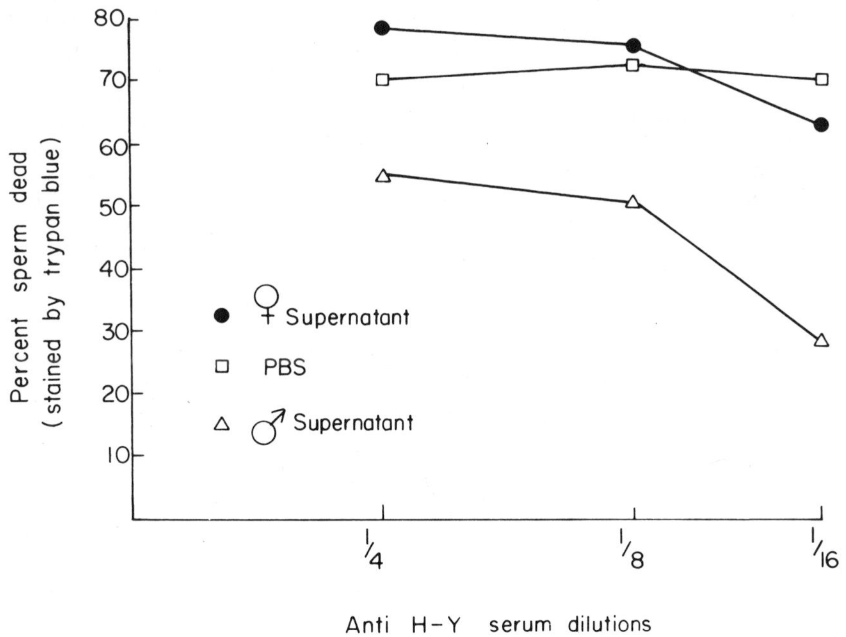

Figure 1: Inhibitory activity of supernatant fluid, in which male lymphocytes have been grown, on sperm cytotoxicity with H-Y antiserum and complement. Serum was diluted in supernatant fluid from an equivalent of 50 x 10^6 male or female lymphocytes/ml.

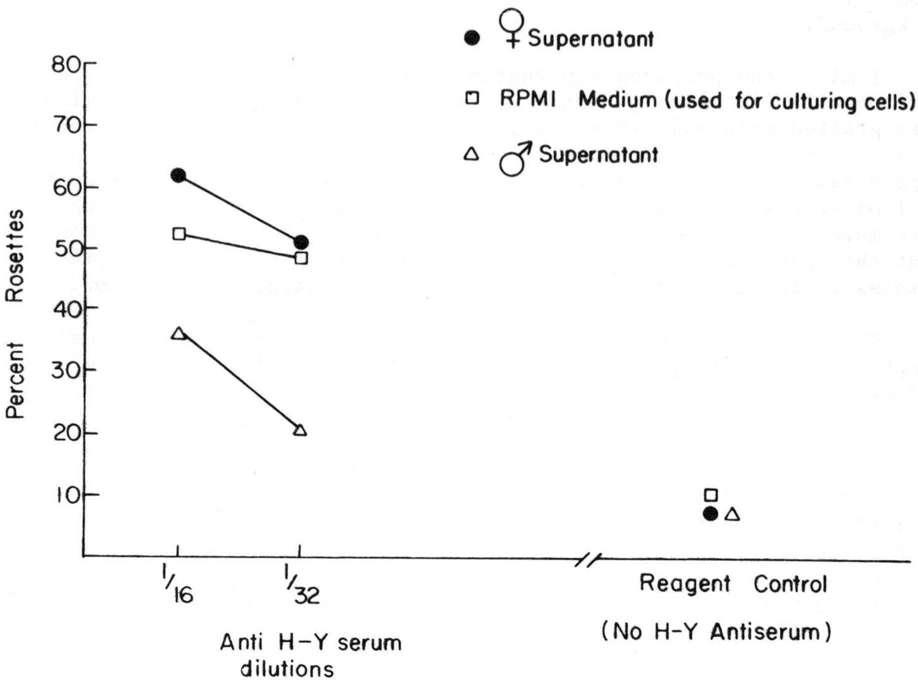

Figure 2: Inhibition of formation of Protein A rosettes on spermatozoa treated with H-Y antiserum diluted in supernatant fluid from cultured male lung fibroblasts.

volumes of supernatant medium from cultured fibroblasts, we chose to use this source of soluble H-Y antigen for further analysis. The supernatant was concentrated 10-fold by an Amicon filtration system with a PM 10 filter which excludes molecules of molecular weight greater than 10,000 daltons. Ammonium sulfate precipitation was carried out at 30, 40 and 50% saturation and it was found that the 40% and 50% precipitates contained the inhibitory activity. For subsequent experiments, therefore, 40%

Table 1: Generation of H-Y specific CTLs by B6 females immunized with supernatant fluid from growth of male lymphocytes.

B6 female immunized with:	Percent specific lysis				
	*3	4	5	6	7
10×10^6 B6 male spleen cells	27	35	22	33	41
B6 male lymphocyte supernatant	28	23	27	32	20
B6 female lymphocyte supernatant	0.2	2	2	1.4	3

*Spleens were removed and the splenocytes put in culture with male stimulator cells 3 to 7 weeks after the primary immunization *in vivo*. Each number represents the average of results from 2 experiments.

Table 2: Rejection of male skin grafts by B6 female mice primed with supernatant fluid from B6 male lymphocytes.

Treatment*	No. of mice	No. of grafts rejected	MST† (days)	Range (days)
Medium (RPMI-1640)	14	14	36.6	(23-45)
B6 female supernatant	14	14	35.2	(22-45)
B6 male supernatant	15	15	24.4	(20-38)

*B6 female mice received 1 ml of either medium alone or supernatant fluid from male or female lymphocytes, 4 weeks before being grafted with B6 male skin.

†MST = mean survival time

ammonium sulfate precipitations were routinely used. After dialysis overnight against phosphate buffered saline (PBS), the precipitate was applied to a Sephadex G-100 (superfine) column with a flow rate of 1 ml/cm^2/2hr. The absorbance at 280 nm of each fraction was measured and 2 major peaks, which corresponded to IgG and albumin (from the fetal calf serum (FCS) present in the medium) were obtained. We tested the inhibitory activity of each fraction, using polyclonal H-Y antiserum and monoclonal H-Y antibody, and found that fractions 40-45 from the male fibroblast supernatant inhibited the H-Y specific reaction of PrA-SRBC rosettes with spermatozoa (Figure 3). When the inhibitory fractions were analysed by 2-dimensional gel electrophoresis, by isoelectric focusing followed by SDS-PAGE, we were able to recognize a series of spots, present in the male supernatant fluid, which were absent from the female supernatant. Preliminary experiments indicate the molecular weight of the male specific material to be 66,000 daltons and its pI to be 5.3. Experiments are currently underway to confirm these findings by immunoaffinity chromatography and Western blotting. At present, we offer no explanation for the discrepancy between our data and that already published, other than the possible influence of the source of material used for isolation and/or the potential for polymorphism of the male specific antigen.

It is clear that the biochemical characterization of H-Y is necessary to help unravel the complexities apparent within the H-Y antigenic system. Questions regarding the multiplicity of H-Y antigens and its potential role in sex differentiation, and the degree of polymorphism and the differences between species should be answerable in the near future as the recent advances in biochemical analyses and gene cloning are exploited. Thirty years ago, when Eichwald and Silmser first described the H-Y antigen, it appeared improbable that it would become such a controversial and potentially important antigenic system. The controversies need to be resolved by collaborations among immunologists, geneticists and molecular biologists. With these kinds of interactions and the availability of new and powerful techniques and exceptional mouse strains, the role played by this fascinating antigenic system in male gonadal differentiation should be established in the very near future.

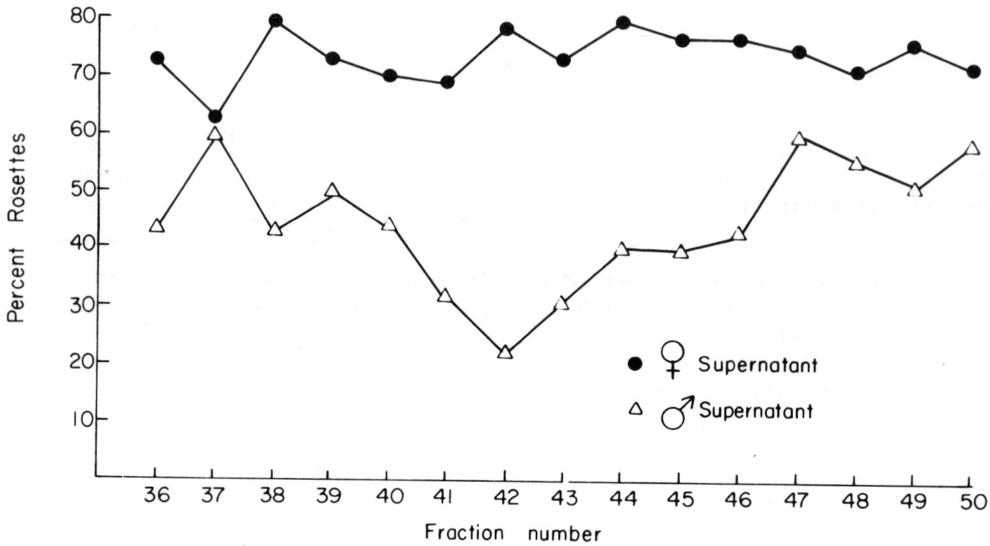

Figure 3: Inhibitory activity of fractions 40-45 on formation of Protein A rosettes, after chromatography of male fibroblast supernatant fluid on a Sephadex G-100 column. Monoclonal H-Y antibody was diluted 1:16 in the fractions before reaction with spermatozoa and Protein A-conjugated sheep red blood cells.

REFERENCES

1. Eichwald, EJ, Silmser, CR: Untitled communication. Transplant Bull, 2:148-49, 1955.

2. Billingham, RE, Silvers, WK: Inbred animals and tissue transplantation immunity with an index to some inbred strains other than mice. Transplant Bull, 6:399-406, 1959.

3. Billingham, RE, Silvers, WK: Induction of tolerance of skin isografts from male donors in female mice. Science, 128:780-81, 1958.

4. Eichwald, EJ, Silmser, CR, Weissman, I: Sex-linked rejection of normal and neoplastic tissue. I. Distribution and specificity. J Natl Can Inst, 20:563-75, 1958.

5. Goldberg, EH, Aoki, T, Boyse, EA, Bennett, D: Detection of H-2 antigens on mouse spermatozoa by the cytotoxicity test. Nature, 228:570-72, 1970.

6. Goldberg, EH, Boyse, EA, Bennett, D, Sheid, M, Carswell, EA: Serological demonstration of H-Y (male) antigen on mouse sperm. Nature, 232:478-80, 1971.

7. Silvers, WK, Yang, SL: Male-specific antigen: Its homology in mice and rats. Science, 181:570-72, 1973.

8. Wachtel, SS, Koo, GC, Boyse, EA: Evolutionary conservation of H-Y ('male') antigen. Nature, 254:270-72, 1975.

9. Krco, CJ, Goldberg, EH: H-Y (male) antigen: Detection on 8-cell mouse embryos. Science, 193:1134-35, 1976.

10. Epstein, CJ, Smith, S, Travis, B: Expression of H-Y antigen on preimplantation mouse embryos. Tissue Antigens, 15:63-7, 1980.

11. Shelton, JA, Goldberg, EH: Male restricted expression of H-Y antigen on preimplantation mouse embryos. Transplantation, 37:7-8, 1984.

12. Ohno, S, Matsunaga, T: The role of H-Y plasma membrane antigen in the evolution of the chromosomal sex determining mechanism. In: Subtelny S, Abbott UK, eds. Levels of Genetic Control. New York: AR Liss, 235-46, 1981.

13. Melvold, RW, Kohn, HI, Yergenian, G, Fawcett, DW: Evidence suggesting the existence of two H-Y antigens in the mouse. Immunogenetics, 5:33-41, 1977.

14. Silvers, WK, Gasser, DL, Eicher, EM: H-Y antigen, serologically detectable male antigen and sex determination. Cell, 28:439-40, 1982.

15. Simpson, E, McLaren, A, Chandler, P: Evidence for two male antigens in mice. Immunogenetics, 15:609-14, 1982.

16. Bailey, DW, Hoste, J: A gene governing the female immune response to the male antigen in mice. Transplantation, 11:404-07, 1971.

17. Gasser, DL, Silvers, WK: Genetic control of the immune response in mice. III. An association between H-2 type and reaction to H-Y. J Immunol, 106:875-76, 1971.

18. Goldberg, EH, Boyse, EA, Scheid, M, Bennett, D: Production of H-Y antibody by female mice that fail to reject male skin. Nature New Biology, 238:55-7, 1972.

19. Goldberg, EH, Goldstein, G, Boyse, EA, Scheid, MP: Effect of the TP5 analogue of thymopoietin on the rejection of male skin by aged and thymectomized female mice. Immunogenetics, 13:201-04, 1981.

20. Coons, TA, Goldberg, EH: Rejection of male skin grafts by splenectomized female mice. Science, 200:320-21, 1978.

21. Goldberg, EH, Shen, FW, Tokuda, S: Detection of H-Y (male) antigen on mouse lymph node cells by the cell to cell cytotoxicity test. Transplantation, 15:334-36, 1973.

22. Gordon, RD, Simpson, E, Samelson, LE: In vitro cell-mediated immune responses to the male specific (H-Y) antigen in mice. Jour Exp Med, 142:1108-20, 1975.

23. Simpson, E, Gordon, RD: Responsiveness to H-Y antigen Ir gene complementation and target cell specificity. Immunol Rev, 35: 59-75, 1977.

24. Mullbacher, A, Brenan M: Cytotoxic T-cell response to H-Y in "nonresponder" CBA mice. Nature, 285:34-6, 1980.

25. Goldberg, EH, Arritt, SE: Effect of thymectomy on the cell-mediated cytotoxicity response to male skin grafts in C3H/HeN female mice. Transplantation, 33:209-10, 1982.

26. Koo, GC, Varano, A: Inhibition of H-Y cell-mediated cytolysis by monoclonal H-Y specific antibody. Immunogenetics, 14:183-88, 1981.

27. Engel, W, Klemme, B, Ebrecht, A: Serological evidence for H-Y antigen in XO-female mice. Hum Genet, 57:68-70, 1981.

28. Koo, GC, Reidy, JA, Nagamine, CM: H-Y antigen in XO mice. Immunogenetics, 18:37-44, 1983.

29. Ohno, S: The Y-linked testis determining gene and H-Y plasma membrane antigen gene: Are they one and the same? Endocrine Review, 6:421-31, 1985.

30. Wachtel, SS, Ohno, S, Koo, GC, Boyse, EA: Possible role of H-Y antigen in primary determination of sex. Nature, 257:235-6, 1975.

31. Erickson, RP: Androgen-modified expression compared with Y linkage of male specific antigen. Nature, 265:59-61, 1977.

32. White, KL, Lindner, GM, Anderson, GB, BonDurant, RH: Cytolytic and fluorescent detection of H-Y antigen on pre-implantation mouse embryos. Theriogenology, 19:701-5, 1983.

33. Ohno, S, Nagai, Y, Ciccarese, S: Testicular cells lysostripped of H-Y antigen organize ovarian follicle-like aggregates. Cytogenet Cell Genet, 20:351-64, 1978.

34. Zenzes, MT, Wolf, U, Gunther, E, Engel, W: Studies on the function of H-Y antigen: Dissociation and reorganization experiments on rat gonadal tissue. Cytogenet Cell Genet, 20:365-72, 1978.

35. Zenzes, MT, Wolf, U, Engel, W: Organization in vitro of ovarian cells into testicular structures. Hum Genet, 44:333-38, 1978.

36. Muller, U, Urban, E: Reaggregation of rat gonadal cells in vitro: experiments on the function of H-Y antigen. Cytogenet Cell Genet, 31:104-7, 1981.

37. Eicher, E: Primary sex determing genes in mice. In: Amann, RP, Seidel, GE, eds. Prospects for Sexing Mammalian Sperm. Boulder: Colo Assoc Univ Press, 121-35, 1982.

38. Washburn, LL, Eicher, EM: Sex reversal in XY mice caused by dominant mutation on chromosome 17. Nature, 303:338-40, 1983.

39. Simpson, E, Chandler, P, Washburn, LL, Bunker, HP, Eicher, EM: H-Y Typing of Karyotypically Abnormal mice. Differentiation, 23:S116-20, 1983.

40. Eicher, EM, Washburn, LL, Whitney, JB III, Morrow, KE: Mus poschiavinus Y-chromosome in the C57BL/6J murine genome causes sex reversal. Science, 217:535-37, 1982.

41. Cattanach, BM, Pollard, CE, Hawkes, SG: Sex-reversed mice: XX and XO males. Cytogenetics, 10:318-37, 1971.

42. Singh, L, Jones, KW: Sex reversal in the mouse (Mus musculus) is caused by a recurrent nonreciprocal crossover involving the X and an aberrant Y chromosome. Cell, 28:205-16, 1982.

43. Bennett, D, Mathieson, BJ, Scheid, M, Yanagisawa, K, Boyse, EA, Wachtel, SS, Cattanach, BM: Serological evidence for H-Y antigen in Sxr, XX sex-reversed phenotypic males. Nature, 265:255-57, 1977.

44. Simpson, E, Edwards, P, Wachtel, SS, McLaren, A, Chandler, P: H-Y antigen in Sxr mice detected by H-2 restricted cytotoxic T cells. Immunogenetics, 13:355-8, 1981.

45. McLaren, A, Simpson, E, Tomonari, K, Chandler, P, Hogg, H: Male sexual differentiation in mice lacking H-Y antigen. Nature, 312:552-5, 1984.

46. Burgoyne, PS, Levy, ER, McLaren, A: Spermatogenic failure in male mice lacking H-Y antigen. Nature, 320:170-72, 1986.

47. Simpson, E, Chandler, P, Hunt, R, Hogg, H, Tomonari, K, McLaren, A: H-Y status of X/X Sxr male mice: in vivo tests. Immunology, 57:345-49, 1986.

48. Nagai, Y, Ciccarise, S, Ohno, S: The identification of human H-Y antigen and testicular transformation induced by its interaction with the receptor site of bovine fetal ovarian cells. Differentiation, 13:155-64, 1979.

49. Ohno, S, Nagai, Y, Ciccarese, S, Iwata, H: Testis-organizing H-Y antigen and the primary sex-determining mechanism of mammals. Rec Prog Horm Res, 35:449-76, 1979.

50. Hall, JL, Bushkin, Y, Wachtel, SS: Immunoprecipitation of human H-Y antigen. Hum Genet, 58:34-6, 1981.

51. Bradley, MP, Heslop, BF: A biochemical and immunological approach to the identification of H-Y antigenic protein secreted from Daudi cells. Hum Genet, 71:117-21, 1985.

52. Shapiro, M, Erickson, RP: Evidence that the serological determinant of H-Y antigen is carbohydrate. Nature, 290:503-5, 1981.

53. Shapiro, M, Goldberg, EH: Analysis of a serological determinant of H-Y Antigen - Evidence for carbohydrate specificity using an H-Y specific monoclonal antibody. J Immunogen, 11:209-18, 1984.

54. Koo, GC, Goldberg, CL: A simplified technique for H-Y typing. J Immunol Methods, 23:197-201, 1978.

SEX DETERMINATION AND H-Y ANTIGEN IN MICE

Anne McLaren

MRC Mammalian Development Unit
Wolfson House, 4 Stephenson Way
London NW1 2HE, England

SUMMARY

DNA sequences in the Sxr region of the mouse Y chromosome play a leading role in primary testis determination. Recent evidence from XX<->XY chimeras suggests that these sequences act cell-autonomously in the supporting cell lineage, but that a Y chromosome-dependent, exogenous signal is also required for the decision to form Sertoli rather than follicle cells. If this signal is given by a diffusible substance, soluble H-Y antigen (as recognized by T lymphocytes) is probably not that substance, but serologically detectable male antigen (if different from H-Y) could be involved.

Sex Determination and H-Y antigen in Mice

Role of the Y chromosome

In mice, as in most other mammals, female individuals are normally XX and males are XY in chromosome constitution. That it is the absence of a Y, rather than the presence of two X chromosomes, that determines femaleness is shown by the fact that the occasional XO individual is female, while XXY and XXXY individuals are male.

However, XX males and XY females have also been reported, albeit rarely. So, does the Y chromosome really play a primary role in sex determination, or is it just one of a number of factors of equal status that interact to produce a male phenotype? The available evidence suggests that the Y chromosome does, indeed, play a primary role. Sex-reversed (Sxr) XX male mice[8] have been shown to carry a piece of the mouse Y chromosome attached to one of their X chromosomes[35,13], while most XX men that have been tested carry Y-specific DNA sequences somewhere in their genome[29]. The more DNA probes for the Y chromosome that are isolated, the more XX men are found to be positive for Y-specific DNA, so it may be that the determination of maleness requires the presence of certain DNA sequences normally associated with the Y chromosome. These hypothetical sequences will be referred to as Tdy (testis-determining sequences on the Y chromosome).

XY females, on the other hand, may arise in a variety of ways, some of which involve the loss of Tdy, while others are consistent with its continued presence, for example:

1) Deletions of the Y chromosome can be detected cytologically in XY women[18,9] and some XY female mice (cited in ref. 10).

2) When the Tdy-carrying region of the mouse Y chromosome (Sxr) is attached to an X chromosome which is inactivated in all cells, the inactivation may spread to include the Tdy sequences and the mouse may then develop as a female[7,23].

3) Although a normal Y chromosome is present, the X chromosome may lack a region that is required to interact with Tdy for the production of maleness. Such is the situation in wood lemmings: an aberrant X, with a cytologically detectable deletion, is common in wild populations, producing a marked skewing of the sex ratio[15,17].

4) XY females have been detected when a Y chromosome (e.g. from a wild population) has been introduced to an evolutionarily remote set of autosomes[11]. This result suggests a failure of certain autosomal sequences (Tda) to respond properly to Tdy.

5) Even if the XY chromosome constitution functions normally and a testis develops, it may lack Leydig cells and, as a result, fail to produce androgens[3,4]. Apart from the gonads and associated ducts, such an individual will have a female phenotype.

6) A similar phenotype is seen in mice and in humans which carry the X-linked mutant gene Tfm (testicular feminization), which renders ineffective all androgen receptors in all target organs[28].

Thus, DNA sequences on the Y chromosome appear to be necessary for the initiation of a cascade of subsequent, male-specific gene expression, both X-chromosomal and autosomal, that results ultimately in the development of a male phenotype. We must now consider how and where this cascade is initiated, in cellular terms.

Development of the testis

As far as we know, the first male-specific cell type to be formed during embryonic development is the pre-Sertoli cell. Within the genital ridge, the early gonadal blastema comprises a homogeneous population of somatic cells, similar in male and female embryos. On the 14th day of gestation in the rat (one day earlier in the mouse, i.e. 12 1/2 days *post coitum*), a new type of cell, characterized by abundant clear cytoplasm, appears in the vicinity of the rete testis[19,1]. These pre-Sertoli cells aggregate together in epithelial-like arrays to form the testis cords, enclosing the great majority of the primordial germ cells. They secrete Mullerian inhibiting substance, causing the primordial female sex ducts to degenerate. The cords become surrounded by peritubular (myoid) cells, and Leydig cells differentiate between them. The pre-Sertoli cells differentiate to form the Sertoli cells that line the seminiferous tubules.

Sertoli cells in the testis are thought to be homologous to follicle cells in the ovary, and Leydig cells to thecal cells. We can, thus, distinguish at least two cell lineages that figure in the development of the gonad (see Figure 1): the supporting cell lineage, which gives rise to Sertoli cells in the male and to follicle cells in the female, and the interstitial cell lineage which differentiates somewhat later. We know that the direction of differentiation of the supporting cell lineage is determined by whether the embryo possesses Tdy sequences. The differentiation of Leydig rather than thecal cells could be dependent on the prior differentiation of Sertoli cells; alternatively, the interstitial cell lineage, too, could respond directly to the presence of Tdy sequences. We may regard the Sertoli cells as primary, not only because they appear earlier in time, but also because testes have been reported which lacked Leydig cells[3,4] but never which lacked Sertoli cells.

Is it a supporting cell's own genes that determine whether it develops as a Sertoli cell or as a follicle cell, or is it the cellular environment (see Figure 2)? In other words, are Tdy sequences expressed cell-autonomously within the supporting cell lineage? Such gene expression could be switched on at any time during the development of this cell lineage or its progenitors, but the most likely time for the initiation of expression of Tdy is, perhaps, at the onset of gonadal differentiation, in response to some non-male-specific environmental trigger. Alternatively, supporting cells could be induced to develop in a male direction, irrespective of whether they themselves possess Tdy sequences, by a signal (for example a diffusible substance) emanating from Tdy-containing cells which could belong to some quite separate lineage.

XX<->XY chimeras usually develop as males[21]. One such male has been reported in which a very high proportion of the cells of the adult gonad belonged to the XX component[25]. The male was sterile and the testes were very small and could, perhaps, have been ovotestes. Nonetheless, the findings suggest that XX cells may be contributing to testicular tissues. It has, therefore, often been assumed that the Y chromosome does not act cell-autonomously in testis determination, but rather that some inducing signal recruits both XY and XX cells to form the somatic cell population of the testis, which includes Sertoli cells[27,38]. However, Burgoyne[5] has pointed out that the evidence for such a view is insubstantial, since determinations of testis

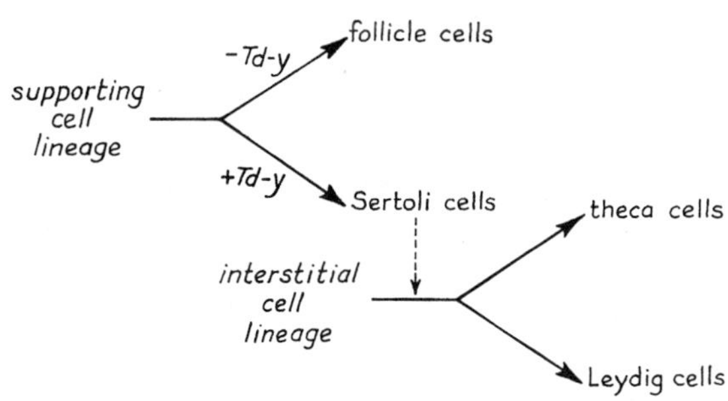

Figure 1: Cell lineage relationships within the genital ridge. The broken arrow represents a hypothetical influence of the Sertoli cells on the direction of subsequent differentiation of the interstitial cell lineage.

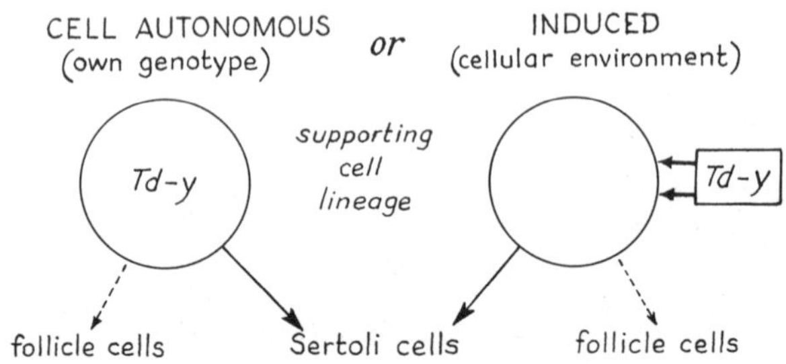

Figure 2: Two alternative ways by which Tdy sequences could influence the supporting cell lineage. Experimental evidence (see text) suggests that both may be operating in normal development of the mouse genital ridge.

composition have all been performed on preparations of whole testis, to which the population of Sertoli cells might contribute a relatively small amount of enzymatic activity. Mouse germ cells located in or near fetal testis cords tend to develop in a male direction, whatever their chromosomal constitution[20], and any that developed as oocytes would be unlikely to survive in the vicinity of testicular tissue[5]. The differentiation of the XY supporting cell component alone, as pre-Sertoli cells, would, therefore, suffice to eliminate virtually all oocytes, with the result that the XX supporting cell component would be unable to differentiate as follicle cells and would be lost.

A chimera experiment

We decided to produce aggregation chimeras between strains distinguished by chromosomal and enzymatic markers, so that we could analyse the contribution of XX and XY cells to an isolated population of Sertoli cells. The results were clear-cut (P.S. Burgoyne, M. Buehr and A. McLaren, unpublished observations). In XY<->XY chimeras, both components contributed to the population of Sertoli cells, as they did to all other somatic tissues analyzed; however in XX<->XY chimeras, the XX component never contributed to the population of Sertoli cells. Both XX and XY components appeared to contribute to the population of Leydig cells, but this result was less decisive, because of contamination of the Leydig cell fraction with other types of cells.

Confirmation of our results comes from a study of chimeras made between *Mus musculus* and *Mus caroli*, using *in situ* hybridization of a radiolabelled, species-specific, DNA probe to tissue sections (J. Rossant, personal communication). In XX<->XY individuals, cells from both components contributed to the population of Leydig cells, but only the XY component was detected in the population of Sertoli cells.

These observations suggest that Tdy is indeed acting cell-autonomously in the supporting cell lineage, to bring about primary testis determination. Subsequent steps in the differentiation of the testis, including determination of the interstitial cell lineage, appear not to be cell-autonomous, perhaps because they are under the control of the supporting cell lineage.

Apart from the presence of Tdy sequences, a further factor seems to be required if the supporting cell lineage is to give rise to Sertoli cells. Female XX<->XY chimeras have been described, in which varying proportions (up to 98%) of the total number of ovarian cells were XY in chromosome constitution[14]. This observation suggests, but does not prove, that XY cells contribute to the population of follicle cells. Our series of chimeras included one XX<->XY individual in which the proportion of XX cells was sufficiently high to allow it to develop as a female. We isolated numerous follicles from the ovaries, and found that the pure population of follicle cells often included a substantial contribution from the XY as well as from the XX component. If follicle cells are, indeed, from the same cell lineage as Sertoli cells, this finding suggests that the presence of Tdy sequences in the genome of a supporting cell is a necessary but not a sufficient condition to determine its development as a Sertoli cell (Table 1). There needs also to be an inducing signal that has to reach a certain threshold intensity, reflecting perhaps a certain minimum ratio of XY to XX cells. This inducing signal, which must itself be controlled by Tdy sequences, could consist of a diffusible substance, but it could equally well involve some factor such as increased growth rate[26], or increased vascular supply. Whatever the nature of the Tdy-controlled signal, it appears that only supporting cells that themselves contain Tdy sequences are capable of responding to it.

One further complication must be noted. Although XX cells cannot give rise to pre-Sertoli cells in the course of primary testis determination, it appears that, in certain circumstances, XX follicle cells can be secondarily transformed into Sertoli-like cells. When fetal mouse ovaries are grafted under the kidney capsule of adult males, the oocytes degenerate and the follicle cells undergo various degrees of transformation, the most extreme product of which closely resembles Sertoli cells, at the ultrastructural as well as at the light microscope level[36]. Similarly, in bovine freemartins, the oocytes degenerate and the follicles eventually transform into cells that resemble Sertoli cells morphologically and also produce Mullerian inhibiting substance, a characteristic product of Sertoli cells[37]. Perhaps, at the time of primary testis determination, the XX supporting cells are too immature to be capable of responding to the Tdy signal; only later do they achieve the competence to respond which the Tdy-containing, XY supporting cells possess at the earlier time. This secondary lability could, perhaps, explain the results of experiments in which differentiated gonads have been disaggregated and exposed to either soluble, male-specific antigen or antibody against such antigen[38].

What is the inducing signal?

It has been suggested that a male-specific antigen, H-Y antigen, acts in mammals as a primary testis-inducing signal. The so-called H-Y hypothesis proposes that somatic cells of both male and female gonads possess receptors for H-Y antigen[38,39], but since H-Y is present only in males, the female gonad differentiates as an ovary. As we have seen, XX cells in the fetal gonad do not appear to respond to the primary testis-inducing signal and do not contribute to the population of Sertoli cells, so the H-Y hypothesis in its original form is probably untenable. However, it is possible that H-Y antigen plays some role in testis induction, either as the Tdy-controlled signal that switches XY supporting cells into the pre-Sertoli cell pathway, or as a secondary signal, produced

Table 1: A possible model for the interaction of endogenous and exogenous Y-associated factors in the differentiation of the supporting cell lineage in the mammalian gonad.

		Supporting cell lineage	
		Tdy present	Tdy absent
Tdy inducing signal	Above threshold	Sertoli cells	Follicle cells
	Below threshold	Follicle cells	Follicle cells

perhaps by pre-Sertoli cells, that switches XX or XY cells of the interstitial lineage into the pathway to Leydig cells.

The male-specific antigen, originally termed H-Y, is a histocompatibility antigen recognized by T lymphocytes, and was first detected by skin-grafting tests[12]. Other T lymphocyte-mediated tests have since heen devised (see ref. 31). With their aid we have shown[33] that X/XSxr males, who have a fragment of the Y chromosome which contains testis-determining sequences attached to one of their X chromosomes, are positive for H-Y antigen, so that genetic information controlling expression of H-Y must be included in the Sxr fragment. We have also shown[34] that Sxr-carrying females (T16H/XSxr) are positive for H-Y antigen. In these females, the presence of an X-autosome translocation T16H ensures that the X chromosome to which Sxr is attached is always inactivated. The proportion of T16H/XSxr embryos that develop as females rather than as males or hermaphrodites varies with the genotype of the father[22]; they are presumed to be mosaics, with one population of cells in which the inactivation process has spread into the Sxr region so as to shut off the Tdy sequences, and another population in which spreading into the Sxr region has not occurred[7,23]. It is, therefore, not surprising that they are positive for H-Y antigen. Their H-Y positive status throws no light on whether or not H-Y is necessary for testis development; it shows only that it is not sufficient.

However, one T16H/XSxr female proved to be H-Y negative[34], and all her male descendants, who were male by virtue of having inherited her Sxr fragment (T16H/XSxr', X/XSxr' and XSxr'/0 males), were also H-Y negative[24]. We presume that this Sxr fragment, which we term Sxr', has been genetically modified, either by point mutation or by loss of some of the DNA, so that it no longer carries the information required for expression of H-Y antigen, as recognized by T lymphocytes. The antigen appears to be totally absent rather than merely altered, since not only do X/X Sxr' males reject skin grafts from XY males[32], but normal XX females accept skin grafts from X/XSxr' males, just as they do from other females (Simpson, E. and McLaren, A., unpublished observations).

Since testis development is identical in X/XSxr and X/XSxr' males, we conclude that the H-Y antigen recognized by T lymphocytes is not necessary for testis development, either as the primary testis inducer or at any later stage. It remains possible (though unlikely) that the H-Y molecule is the testis inducer, and that in Sxr' it has lost its antigenicity while retaining its biological function.

A possible alternative function for H-Y antigen was suggested by a comparison of XSxr/0 and XSxr'/0 males. The testes of XSxr/0 males show active spermatogenesis (unlike those of X/XSxr males, in which all germ cells degenerate shortly after birth as a result of the presence of two X chromosomes). Spermatogonia are present in abundance, and all stages of spermatogenesis are seen, though the few spermatozoa that are formed are abnormal and largely immotile. The testes of XSxr'/0 males, in contrast, show virtually no spermatogenesis[6]. They contain only a few scattered spermatogonia and primary spermatocytes up to the pachytene stage. It appears that the transition from Sxr to Sxr' has involved the loss of some genetic information required for spermatogenesis. If the transition was due to a point mutation, H-Y antigen itself must play a role in spermatogenesis. If, as seems more likely, the transition was due to loss of chromosomal material, the locus involved in spermatogenesis could be separate from the H-Y locus, but closely linked to it.

When female mice are immunized with male tissue, their B lymphocytes respond by producing specific anti-male antibodies[16]. The male-specific antigen responsible for this serological response was assumed to be the same as that recognized by T lymphocytes, so it too was referred to as H-Y antigen. Since there is no conclusive evidence that the two antigens are the same or that they are different, the antigen recognized by B lymphocytes is now, for the sake of clarity, often referred to as SDM (serologically detectable male) antigen[30]. The H-Y hypothesis[39] did not distinguish between the two antigens, and, indeed, much of the subsequent work exploring the hypothesis was actually concerned with SDM rather than with H-Y. If the two antigens are different, it remains quite possible that SDM functions either as the Tdy-controlled signal that switches XY supporting cells into the Sertoli cell pathway, or as the secondary signal that switches XX or XY interstitial cells into the Leydig cell pathway. Unfortunately SDM antigen is fragile, antisera to it are usually of low titre, and the serological assays are hard to establish and sometimes unreliable. X/XSxr males are known to be positive for SDM antigen[2], but it has not yet proved possible to type X/XSxr' males.

CONCLUSIONS

The Sxr chromosome fragment is large enough to be visualized cytologically, but is probably not more than a few thousand kilobases in length. It contains genetic information of crucial importance both for testis determination and for spermatogenesis, and controls the expression of one, or perhaps two, male-specific antigens. Thus, we know of four possible genetic loci, structural or regulatory, within this region. However, we cannot assume that these are 4 separate loci: Table 2 indicates some possibilities in which the data are explained by 3 loci, 2 loci or even just one locus. The last possibility assumes a degree of evolutionary thrift that seems biologically implausible. Alternatives (c) and (e), either of which would require two loci, seem to combine biological plausibility with an appropriately vigorous use of Occam's famous razor.

Table 2: Four hypothetical loci within the Sxr fragment, and their possible relationships

Suggested name of locus	Associated gene product	Biological function
Tdy	?	Testis determination
Spy	?	Spermatogenesis
Hya	Male antigen (transplantation)	?
Sdm	Male antigen	

Hypotheses:		
	a) Sdm = Tdy	(3 loci)
	b) Hya = Spy	(3 loci)
	c) Sdm = Tdy, Hya = Spy	(2 loci)
	d) Sdm = Hya	(3 loci)
	e) Sdm = Hya = Spy	(2 loci)
	f) Sdm = Hya = Spy = Tdy	(1 locus)

REFERENCES

1. Agelopolou, R, Magre, S, Patsavoudi, E, Jost, A: Initial phases of the rat testis differentiation in vitro. J Embryol Exp Morph, 83: 15-31, 1984.

2. Bennett, D, Mathieson, BJ, Scheid, M, Yanagisawa, K, Boyse, EA, Wachtel, S, Cattanach, BM: Serological evidence for H-Y antigen in Sxr, sex-reversed phenotypic males. Nature, 265:255-257, 1977.

3. Berthezene, F, Forest, MG, Grimaud, JA, Clanstrat, B, Mornex, R: Leydig cell agenesis : a cause of male pseudohermaphroditism. New Engl J Med, 259:969-972, 1976.

4. Brown, DM, Markland, C, Delmer, LP: Leydig cell hypoplasia: a cause of male pseudohermaphroditism. J Clin Endocrinol Metab, 46:1-7, 1978.

5. Burgoyne, PS, Ansell, JD, Tournay, A: Can the indifferent mammalian XX gonad be sex-reversed by interaction with testicular tissue? In: Eshkol A, ed. Serono Symp. Development and Function of the Reproductive Organs, London: Academic Press, 1986; 11:23-39.

6. Burgoyne, PS, Levy, ER, McLaren, A: Spermatogenic failure in male mice lacking H-Y antigen. Nature, 320:170-172, 1986.

7. Cattanach, BM, Evans, EP, Burtenshaw, M, Barlow, J: Male, female and intersex development in mice of identical chromosome constitution. Nature, 300:445-446, 1982.

8. Cattanach, BM, Pollard, CE, Hawkes, SG: Sex-reversed mice: XX and XO males. Cytogenetics, 10:318-337, 1971.

9. Disteche, CM, Brown, L, Saal, H, Friedman, C, Sybert, V, Graham, J, Thuline, H, Page, DC, Fellous, M: Small deletions of the short arm of the Y chromosome in 46,XY females. Proc Natl Acad Sci USA, 1986; in press.

10. Eicher, EM: Primary sex determining genes in mice. In: Amann RP, Seidel GE, eds. Prospects for sexing mammalian sperm. Boulder: Colorado Assoc. Univ. Press, 121-135, 1982.

11. Eicher, EM, Washburn, LL, Whitney, JB, Morrow, KE: Mus poschiavinus Y chromosome in the C57BL/6J murine genome causes sex reversal. Science, 217:535-537, 1982.

12. Eichwald, EJ, Silmser, CR: Untitled communication. Transplant Bull, 2:148-149, 1955.

13. Evans, EP, Burtenshaw, M, Cattanach, BM: Cytological evidence for meiotic crossing-over between the X and Y chromosome of male mice carrying the sex reversing (Sxr) factor. Nature, 300:443-445, 1982.

14. Ford, CE, Evans, EP, Burtenshaw, MD, Clegg, H, Barnes, RD, Tuffrey, M: Marker chromosome analysis of tetraparental AKR CBA-T6 mouse chimaeras. Differentiation, 2:321-333, 1974.

15. Fredga, K, Gropp, A, Winking, H, Frank, F: Fertile XX- and XY-type females in the wood lemming (Myopus schistocolor). Nature, 261:225-227, 1976.

16. Goldberg, E, Boyse, EA, Bennett, D, Scheid, M, Carswell, EA: Serological demonstration of H-Y (male) antigen on mouse sperm. Nature, 232:478-480, 1971.

17. Herbst, EW, Fredga, K, Frank, F, Winking, H, Gropp, A: Cytological identification of two X-chromosome types in the wood lemming (Myopus schistocolor). Chromosoma, 69:185-191, 1978.

18. Jacobs, PA, Ross, A: Structural abnormalities of the Y chromosome in man. Nature, 210:352-354, 1966.

19. Jost, A: Donnees preliminaires sur les stades initiaux de la differenciation du testicule chez le rat. Arch Anat Micr Morph Exper, 61:415-438, 1972.

20. McLaren, A: Meiosis and differentiation of mouse germ cells. In: Evans CW, Dickinson HG, eds. Controlling Events in Meiosis, Symp Soc Exp Biol, 38:7-23, 1984.

21. McLaren, A: Chimeras and sexual differentiation. In: Le Douarin N, McLaren A, eds. Chimeras in Developmental Biology. London: Academic Press 381-399, 1984.

22. McLaren, A: Sex ratio and testis size in mice carrying Sxr and T(X;16)16H. Dev Genet, 6, 1986.

23. McLaren, A, Monk, M: Fertile females produced by inactivation of an X chromosome of sex-reversed mice. Nature, 300:446-448, 1982.

24. McLaren, A, Simpson, E, Tomonari, K, Chandler, P, Hogg, H: Male sexual differentiation in mice lacking H-Y antigen. Nature, 312: 552-555, 1984.

25. Mintz, B: Developmental mechanisms found in allophenic mice with sex chromosomal and pigmentary mosaicism. In: Bergsma D, McKusick V, eds. 1st Conf. on the Clinical Delineation of Birth Defects. Original Article Series 5, Birth Defects. New York: National Foundation, 11-22, 1969.

26. Mittwoch, U: Heterogametic sex chromosomes and the development of the dominant gonad in vertebrates. Amer Nat, 122:159-180, 1983.

27. Ohno, S, Ciccarese, S, Nagai, Y, Wachtel, SS: H-Y antigen in testes of XX(BALB)--XY(C3H) chimaeric male mouse. Arch Androl, 1:103-109, 1978.

28. Ohno, S, Lyon, MF: X-linked testicular feminization in the mouse as a non-inducible regulatory mutation of the Jacob-Monod type. Clin Genet, 1:121-127, 1970.

29. Page, DC: Sex reversal: deletion mapping the male-determing function of the human Y chromosome. In: Molecular Biology of Homo sapiens. Cold Spring Harbor Symp. Quant. Biol. 51:1986, in press.

30. Silvers, WK, Gasser, DL, Eicher, EM: H-Y antigen, serologically detectable male antigen and sex determination. Cell, 28:439-440, 1982.

31. Simpson, E. Immunology of H-Y antigen and its role in sex determination. Proc R Soc Lond B, 220:31-46, 1983.

32. Simpson, E, Chandler, P, Hunt, R, Hogg, H, Tomonari, K, McLaren, A: H-Y status of X/X Sxr' male mice: in vivo tests. Immunology, 57:345-349, 1986.

33. Simpson, E, Edwards, P, Wachtel, S, McLaren, A, Chandler, P: H-Y antigen in Sxr mice detected by H-2 restricted cytotoxic T cells. Immunogenetics, 13:355-358, 1981.

34. Simpson, E, McLaren, A, Chandler, P, Tomonari, K: Expression of H-Y antigen by female mice carrying Sxr. Transplantation, 37:17-21, 1984.

35. Singh, L, Jones, KW: Sex reversal in the mouse (Mus musculus) is caused by a recurrent nonreciprocal crossover involving the X and an aberrant Y chromosome. Cell, 28:205-216, 1982.

36. Taketo-Hosotani, T, Merchant-Larios, H, Thau, RB, Koide, SS: Testicular cell differentiation in fetal mouse ovaries following transplantation into adult male mice. J Exp Zool, 236:229-237, 1985.

37. Vigier, B, Tran, D, Legeai, L, Bezard, J, Josso, N: Origin of anti-Mullerian hormone in bovine freemartin fetuses. J Reprod Fert, 70:473-479, 1984.

38. Wachtel, SS: H-Y antigen and the biology of sex determination. London: Academic Press, 1982.

39. Wachtel, SS, Ohno, S, Koo, GC, Boyse, EA: Possible role for H-Y antigen in the primary determination of sex. Nature, 257:235-236, 1975.

40. Washburn, LL, Eicher, EM: Sex reversal in XY mice caused by dominant mutation on chromosome 17. Nature, 303:338, 1983.

GENES ON THE HUMAN Y CHROMOSOME

P.N. Goodfellow, P.J. Goodfellow, B. Pym, G. Banting
C. Pritchard and S.M. Darling

Laboratory of Human Molecular Genetics
Imperial Cancer Research Fund
Lincoln's Inn Fields, London, WC2A 5PX, U.K.

INTRODUCTION

The human Y chromosome is remarkable not only for the genes it encodes, but, also for the genes it does not encode. Unique among the human chromosomes, the Y chromosome is apparently deficient in genetic information. At a recent official count[1], over 100 genes were assigned to the human X chromosome. As the Y chromosome is approximately one third the size of the X, it might be expected that 30 genes would have been mapped to the Y. Even if all the "may be" and "perhaps" assignments are included, only 7 genes have been mapped to the Y chromosome and, of these, only a single gene has passed the final test of being cloned and subjected to sequence analysis[2,3].

The absence of genetic information from the Y chromosome is a direct consequence of its function and of dosage compensation by X-inactivation in mammals. The evolution of the relationship between sex determination and X-inactivation is not clear[4,5,6]. However, this relationship has resulted in the Y chromosome of man having two distinct parts whose characteristics must meet different structural requirements.

The Y chromosome must pair with the X chromosome and go to a different pole during male meiosis. This function is achieved by the sharing of sequences by the X and Y chromosomes, and recombination between these sequences during male meiosis ensures the identity which is needed for meiotic pairing[7,8,9]. The part of the sex chromosomes in which sequences recombine has been termed the pseudoautosomal region (see below). Theoretically, any gene could be located in the pseudo-autosomal region, as it would be present identically in males and females. This assumption presupposes that the X-located gene escapes X-inactivation. The mammalian pseudoautosomal genes, so far described, escape X-inactivation[10,11].

The second functional component of the Y chromosome encodes *TDF*, the gene (or genes) required for sex determination. *TDF* must not recombine with the X chromosome, since such recombination would lead to a breakdown in the chromosomal mechanism of sex determination. The phenotypic consequences of transfer of *TDF* to the X chromosome can be seen in *XX* males[12,13]. The segment of the *Y* chromosome which carries *TDF* would *not* be expected to share genes or sequences with the X chromosome, because such sequence homology might promote recombination. The absence of genes from this region can also be explained by X-inactivation. A gene, outside the pseudoautosomal region of the Y chromosome, which is required in both males and females, must be present elsewhere in the genome. For reasons of dosage the only likely, extra position for the homolog of a Y-linked gene would be the X chromosome. However, the X chromosome has evolved a dosage compensation mechanism and any gene which is X-inactivated in females is unlikely to be present on the Y chromosome. Recent molecular analysis has, paradoxically, demonstrated extensive sequence homology not only between the Y and the X chromosome, but also between the Y chromosome and autosomes (reviewed in ref. 14). These homologies are due to recent transpositions in the primate lineage and, presumably, are found in positional arrangements on the Y chromosome which do not promote chromosomal recombination.

Two classes of genes may be present on the Y-specific part of the Y chromosome. The first class includes genes which are required for the male phenotype and are harmful to the female phenotype. Genes involved in spermatogenesis may fall into this class, and it has been proposed that the requirement for more than one Y-specific gene for development of the full male phenotype was the driving force behind the isolation of the Y chromosome from the X chromosome[4].

The second class of genes which might be present on the Y chromosome would include those genes expressed during early embryogenesis, before X-inactivation. After X-inactivation, the Y-located gene would also need to be inactivated. There is currently no direct evidence for such genes; X-linked genes, known to be expressed during early mouse embryogenesis show the expected difference in dosage between male and female embryos (see, for example, ref. 15).

In this review of genes on the human Y chromosome, we concentrate unashamedly on our own work. Readers requiring more background information about the Y chromosome are directed to several recent reviews[14,16,17] and several of the topics discussed briefly in this chapter are treated in depth elsewhere in this volume. Because of the differences between the pseudoautosomal region and the rest of the Y chromosome, we have divided the review accordingly.

Pseudoautosomal genes

Molecular biology has succeeded in confirming a prediction made by Darlington half a century ago. The human sex chromosomes share a region of identical sequences which can recombine during meiosis[18,19,20,21]. Sequences which recombine between the sex chromosomes show an apparent autosomal pattern of inheritance, and such sequences have been termed pseudoautosomal. Several indirect arguments suggest that the pseudoautosomal region consists of about 3×10^6 bp[22]. Recombination in males in this region is remarkable because the genetics suggest a single, obligate exchange at each meiotic event in males. This result means that 50% recombination occurs in fewer than 5×10^6 bp of DNA, and this value implies a rate which is 10 times higher than the average for the whole genome (see ref. 23). The biological consequences of this high rate of recombination are not known. However, it may not be a coincidence that the pseudoautosomal region, unlike the rest of the X chromosome, shows very high levels of polymorphism, as deduced from restriction enzyme fragment length polymorphism (RFLP) analysis.

In the mouse, the gene STS, encoding the enzyme steroid sulphatase, is pseudoautosomal[10]. The genetics of the equivalent locus STS in man is only consistent with classical X-linkage[24]. In situ hybridization with molecular probes derived from the human STS gene have confirmed the X-location and have excluded the possibility of a Y-located, pseudoautosomal STS gene in man[25]. A sequence distantly related to STS is present on the long arm of the Y chromosome (ref. 25; and L. Shapiro, personal communication).

In man, only the gene MIC2 has been formally shown to be pseudoautosomal. However, in this chapter we will discuss some recent results which have led us to postulate that a second gene, XGR, is also pseudoautosomal. XGR is a cis acting locus which regulates expression of both the XG antigen and the 12E7 antigen on red blood cells.

The MIC2 gene

The MIC2 gene encodes a cell-surface antigen which is recognised by the monoclonal antibody 12E7. The antigen is expressed on most human cells, but is not present on the surface of rodent cells. Human-rodent somatic cell hybrids which retain either the human X or the Y chromosome express the 12E7 cell-surface antigen. These experimental results define the loci MIC2X and MIC2Y. In addition, human-mouse hybrids which retain an inactive human X chromosome also express the 12E7 antigen, demonstrating that the MIC2X locus escapes X-inactivation[11]. Thus, gene dosage and expression of MIC2 is maintained between males and females. The genetics of the MIC2 system and our interest in the human sex chromosomes prompted us to investigate the biochemistry and molecular biology of the MIC2 locus; in particular we wanted to know the extent of the relationship between MIC2X and MIC2Y.

Western blot analysis, using the 12E7 antibody, has shown that the product of the MIC2 gene is associated with a molecule of molecular weight of 32,000 daltons. The antibody also recognises a cytoplasmic molecule of molecular weight of 29,000 daltons, but this smaller molecule is found in both mouse and human cells. No differences in size or charge could be detected between the cell-surface products derived from either the MIC2X or the MIC2Y locus[28]. Based on our ability to detect the MIC2 products using Western blot analysis, we isolated cDNA clones that encode the 12E7 antigen, using a bacterial expression system[2]. The clones detected identical restriction fragments in Southern blot analysis of genomic DNA from males and females. In situ hybridization, using the MIC2 cDNA clone, confirmed the localisation of MIC2X to Xp22.32-pter and refined the localisation of MIC2Y to Yp11-pter. These locations are consistent with the postulated pairing region[29].

Final proof that MIC2X and MIC2Y are both identical and pseudoautosomal was obtained from familial studies, using cDNA and genomic probes which recognised RFLPs. MIC2 recombines between the X and Y chromosome with a frequency of a few percent[21], so that MIC2 is the most proximal marker in the pseudoautosomal region and the closest known distal marker to TDF. We are currently using chromosome-mediated gene transfer, in association with selection based on the presence of MIC2, to isolate fragments of the Y chromosome which include TDF[30].

The biochemical function of the MIC2 product is not known. Sequence analysis has shown that the product is a transmembrane protein with an external C-terminus[3]. No significant homology has been detected with any other protein that has been characterised thus far.

Genes regulating levels of expression of the 12E7 antigen on red blood cells

With the possible exception of spermatozoa, the 12E7 antigen is expressed on all human cells, including red blood cells[31]. The levels of expression of 12E7 antigen on cultured skin fibroblasts and peripheral blood lymphocytes do not vary between cells derived from different individuals. In contrast, expression of 12E7 antigen on red blood cells is polymorphic[52]. Two phenotypes can be distinguished: low level expression and high level expression. In quantitative assays, red cells with the "low level antigen" phenotype bind approximately one third the amount of 12E7 antibody as is bound by red cells of the "high level antigen" phenotype. This polymorphism is also detected by other monoclonal antibodies which react with the 12E7 antigen (unpublished observations). A complex relationship exists between the genetic control of the polymorphism of the 12E7 antigen and the XG locus[32].

Classically, XG has been described as the X-linked, structural locus for the red blood cell antigen XG. This locus is polymorphic with two alleles XG*A and XG*B (our "red blood cell" friends will be outraged by this nomenclature, which is recommended by the human geneticists, see ref. 33; in "red blood cell" terminology these alleles would be written Xg^a and Xg). The XG*A antigen is defined by reaction with XG antisera. XG*B is the reciprocal "null" allele. The XG locus is located close to MIC2 in Xp22.3 and escapes X-inactivation[34,35]. Extensive familial and population studies have ruled out the possibility of a Y-located form of XG[35].

In females, the phenotype of the XG antigen predicts the phenotype of the 12E7 antigen. XG antigen-positive females (XG*A/XG*A; XG*A/XG*B) express high levels of 12E7 antigen; antigen-negative females (XG*B/XG*B) express low levels of 12E7 antigen. In males, further complexity is encountered: all XG antigen-positive males express high levels of 12E7 antigen; however, antigen-negative males can express either low levels or high levels of 12E7 antigen. To explain these data, we postulated the existence of a polymorphic, Y-located locus, YG, which has the property of regulating expression of 12E7 antigen on red blood cells. The YG*A allele would cause high level expression of 12E7; the YG*B allele would cause low level expression. The existence of YG and its Y-location has been proved by a combination of familial studies and an ingenious use of sibling analysis[36].

Those readers who have managed to follow the argument so far will not be surprised to learn that recent studies have suggested that the above description of the interaction between the pseudoautosomal locus MIC2 and the X-located XG and the Y-located YG is an oversimplification. By use of cloned probes for MIC2 as markers in familial studies, MIC2 has been shown to be pseudoautosomal[21] and, as expected, MIC2 recombines with XG (unpublished results). This result formally distinguishes between XG and MIC2. Although the overwhelming evidence from analysis of siblings showed that XG antigen-negative brothers shared the same 12E7 antigen phenotype, 6 exceptional cases were found in the 200 brothers tested[36]. The majority of these exceptions was probably caused by technical difficulties; testing for the presence of 12E7 antigen on old samples of red cells is not reliable. However, one exceptional case was investigated in detail and found to be the result of an apparent recombination between the X and Y chromosome. An XG*B, YG*A father had a son who apparently had the genotype XG*B, YG*B[36]. This case has been investigated with molecular probes and the son has been found to be recombinant also at MIC2. Recombination between the X and Y chromosomes at MIC2 occurs with a frequency of only a few percent and the possibility that two independant, rare events have occured in this individual is unlikely. In order to accommodate these new data we propose:

1. XGS is an X-located, structural locus for the XG antigen. XGS is not polymorphic and is not pseudoautosomal.

2. XGR is a pseudoautosomal, regulatory locus which has two alleles: XGR*A and XGR*B. The *A allele works in cis to induce the XGS locus to produce the XG antigen and to induce the MIC2 locus to produce high levels of the 12E7 antigen on red cells. The *B allele induces the XG antigen-negative phenotype with the phenotype of the low level of expression of 12E7 antigen. These relationships are described in detail in Table 1.

3. MIC2 is the structural locus for the 12E7 antigen.

4. The order of the loci on the X chromosome is centromere-XGS-XGR-MIC2-telomere (see Figure 1).

This new model describes the available data and predicts that the frequency of expressors of high and low levels of 12E7 will be the same as the frequency of the XG antigen-positive and -negative phenotypes. This prediction is based in part on mathematical modelling which suggests that very low levels of recombination (of the same order of magnitude as the mutation rate) between the sex chromosomes will maintain identical frequencies of alleles in the pseudoautosomal region[9]. Available data support the prediction[32].

Table 1: Genotype-phenotype relationships which result from polymorphism at XGR

GENOTYPES	PHENOTYPE		
	SEX	XG ANTIGEN	12E7 ANTIGEN
XGS. XG*A. MIC2 XGS. XG*A (OR *B) MIC2	F	+	HIGH
XGS. XG*B. MIC2 XGS. XG*B. MIC2	F	-	LOW
XGS. XG*A. MIC2 TDF. XG*A (OR *B) MIC2	M	+	HIGH
XGS. XG*B. MIC2 TDF. XG*A. MIC2	M	-	HIGH
XGS. XG*B. MIC2 TDF. XG*B. MIC2	M	-	LOW

Several components of the model have not been formally demonstrated. For example, we have not shown directly that MIC2 encodes the red blood cell form of the 12E7 antigen. Finally, the model makes the unlikely suggestion that a *cis* acting element can operate on two separate genes.

Y-linked genes which are not pseudoautosomal

In theory, the determination of Y-linkage should be simple: individuals sharing the same Y chromosome should share the same Y genes and express the same phenotype. In practice, demonstration of Y-linkage may be hindered by sex-limitation. Any phenotype which is only expressed in males or is affected by the male phenotype may have the appearance of Y-linkage, especially if only small numbers of families are available for study. Another problem is that most of the candidates for Y-location are genes which affect complex phenotypes, such as size and growth. Furthermore, studies of quantitative genetics in an outbred species, such as man, are difficult. Three classes of strictly Y-linked genes have been described in man: sex determining genes, H-Y antigen genes, and genes which regulate complex phenotypes. Several pseudogenes have also been assigned to the Y chromosome.

Sex determining genes

Undoubtedly the most interesting gene (or genes) on the human Y chromosome is the sex determining gene. Unfortunately, our interest is matched only by our ignorance.

If sex determination occurs in humans in a similar fashion to that in other mammals, then the presence of the Y chromosome generates a

Figure 1: A Genetic Model to Explain the Relationship Between the XG and 12E7 Antigen Phenotypes on Red Blood Cells

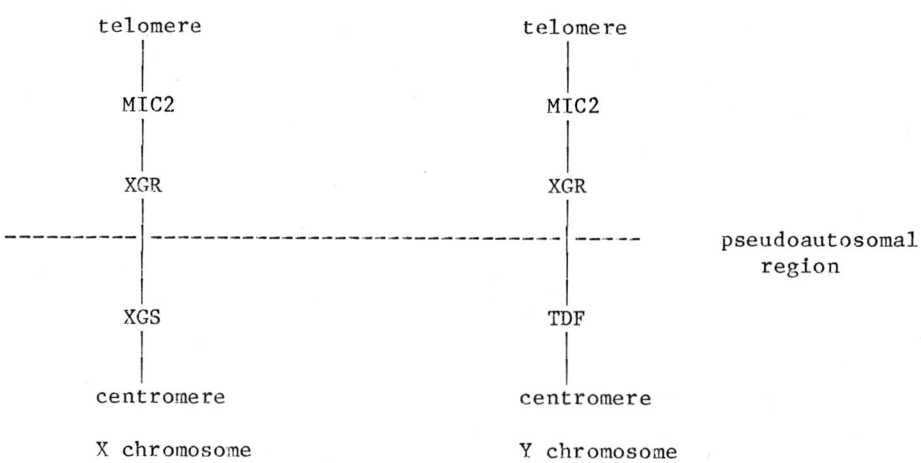

MIC2 : structural gene for the 12E7 antigen
XGS : structural gene for the XG antigen
XGR : regulator locus which affects XGS and MIC2 in *cis*
TDF : the testis determining factor

signal which induces testicular differentiation in the primative gonad[37]. Experiments in mice, using chimeras composed of XX and XY cells, strongly implicate the sertoli cells as the site of action of this signal (ref. 38; P.Burgoyne, personal communication). The nature of the primary signal and the number of Y-located and other genes which are needed to generate the signal are unknown. It is not even known if the sex determining sequences encode polypeptides or work indirectly by binding "repressors"[39].

In human genetics, the lack of information about the sex determining gene has been recognised in the use of the term "testis determining factor", TDF. Note that factor is the vaguest term that still has genetic meaning. Although little is known of the biology of TDF, recent molecular analysis of the Y chromosome, and of the Y-derived sequences present in XX males, has simplified the genetics and located TDF on the short arm of the Y chromosome. These data are reviewed in detail elsewhere in this volume. In brief, many XX males contain varying amounts of detectable, Y-derived sequences. Assuming that most XX males are generated by a single abnormal exchange between the terminal short arms of the X and Y chromosomes, then a linear and largely consistent map of the Y chromosome can be constructed. This map places TDF just below MIC2 and the rest of the pseudoautosomal region[21]. On theoretical grounds, we have suggested elsewhere that TDF may even delimit the pseudoautosomal region[9]. These studies are consistent with TDF mapping to a single region and they exclude the possibility of multiple, additive factors scattered throughout the Y chromosome. The majority of the data is consistent with a single map order for the Y chromosome. However, polymorphism for large inversions has been suggested[13,40]. Many of the problems that remain could be resolved by exchange of reagents and biological material between the laboratories that are active in this field.

Now that the TDF gene has been located, it is hoped that this map position can be used as a basis for molecular cloning. Perhaps the secrets of TDF will soon be revealed.

H-Y antigen genes

Several books have been written about the H-Y antigen (e.g. refs. 4 and 41) and further reviews can be found in this volume. Obviously we cannot match this depth of treatment and, in consequence, we will restrict ourselves to the expression of a few personal prejudices.

H-Y antigen was originally discovered by skin grafting between male and female inbred mice: males accepted female skin, but females rejected male skin. This response defined a minor histocompability antigen which was dependent on the presence of the Y chromosome. Subsequent experiments in mice, using transplantation or the analogous assay *in vitro* of T cell killing, demonstrated that the responsible locus was closely associated with the sex determining gene (reviewed in ref. 42), but distinct from it[43].

Biochemical studies are very difficult to perform when T cell killing or transplantation are used as assays for antigens. For this reason, considerable effort has been directed towards making antisera or monoclonal antibodies which react with the H-Y antigen. Unfortunately, it is very difficult to make and assay antibodies to minor histocompatibility antigens, perhaps because these antigens are cytoplasmic proteins and must be presented at the cell surface by class 1 major histocompatibility antigens[44]. The serological problems have resulted in a large and often conflicting literature on H-Y antigens, as defined by serology. These problems have also been confounded by the apparent absence of polymorphism detected by the different reagents. A polymorphic difference could be used to show formally the Y-linkage of the gene (or genes) that control expression of serologically detectable H-Y antigens in males. At present, the unexpected discovery of the antigen in females may be explained by undetected, Y-derived sequences, chimerism, or the location of the structural gene (or genes) on chromosomes other than Y. Similar arguments can be made for unexpected absence of the antigen from males.

The serological problems have prevented evaluation of several other crucial questions: what is the relationship between serological and transplantation H-Y antigens; how many different H-Y antigens are detected serologically; and what are the biochemical characteristics of H-Y antigens? Recently described, high titre H-Y antisera may resolve some of these difficulties[45].

Most studies of H-Y antigen in man have depended on contentious serology. However, T-cell killing assays can be used to define H-Y in humans[46]. Unpublished studies using T cell killing assays have failed to detect expression of H-Y antigen in human XX males with Y-derived sequences (E.Simpson, personal communication). Thus, evidence from both mouse and man suggests that the H-Y gene (as defined by transplantation) is distinct from the sex determining genes. The function of the H-Y antigen is unknown. However, Burgoyne et al[47] have presented evidence that H-Y antigen may affect spermatogenesis. If the gene controlling expression of expression of human H-Y antigen maps in the same region as the factors which affect spermatogenesis in man (see below), this suggestion would be supported.

Genes which regulate complex phenotypes

The difficulties associated with the study of the genes which regulate complex phenotypes have been alluded to in the introduction to this section.

Perhaps the most convincing evidence has been presented for a gene which controls the bizarre phenotype of extremely hairy ears[48]. Unfortunately, the phenotype appears to be found only in the Indian subcontinent and consequently does not provide a general marker for studies of the Y-chromosome. No information about the location on the Y chromosome has been presented for this gene.

Analysis of patients with deletions of the long arm of the Y chromosome has been used to define AZF, a factor needed for spermatogenesis. Azoospermia (absence of sperm production) is a general feature which results from deletion of the proximal region of the long arm, but alternative explanations, such as a minimum chromosomal size required for normal meiotic segregation, have not been excluded[49,16]. If the gene controlling expression of the H-Y antigen (as defined by T cell killing) maps to this same region, it would strengthen the suggestion that H-Y antigen is involved in spermatogenesis.

Without reliance solely upon familial studies or analyses of abnormal sex chromosomes, a gene (or genes) affecting growth has been assigned to the Y chromosome, based on dosage. XYY males have larger teeth than XY men, who in turn have larger teeth than XX males[50]. Obviously such arguments cannot easily be used to exclude indirect effects which may contribute to the observed phenotype.

Y-located pseudogenes

The definition of a pseudogene has become confused in the literature. A pseudogene is a gene which cannot be expressed to produce a functional product. However, pseudogenes may compete at the DNA level for regulatory molecules and, as such, may not be totally devoid of function. Recent fashion has been to describe any intronless, presumably RNA-derived sequence as a pseudogene. Such an assumption is dangerous, some of these genes may be expressed. Even more dangerous is the assumption that a band on a Southern blot, which reacts only weakly with a cDNA probe, is a pseudogene. Obviously many other explanations are possible.

With these strictures in mind, we note that one actin pseudogene *ACTP1* and two argininosuccinase pseudogenes *ASSP4* and *ASSP5* have been mapped on the Y chromosome. The actin pseudogene was defined by an α-actin probe and subsequent cloning and sequencing of the Y-located actin gene confirmed its inability to be expressed[51]. Similar data has been presented for *ASSP4* and *ASSP5*[17,52].

Given the extensive sequence homologies found between the autosomes and the Y chromosome, it would not be surprising if many other pseudogenes were located on the Y chromosome.

CONCLUSIONS

The Y chromosome is strikingly deficient in known, expressed, genetic material. This deficiency could reflect a bias based on analytical procedures. However, a more likely explanation is the absence of genes from the Y chromosome. Apart from *TDF*, and perhaps a few genes which contribute to the male phenotype, we think it is unlikely that other expressed genes will be found on the Y-specific part of the Y chromosome. This absence of expressed genes is a direct consequence of the chromosomal basis of sex determination and the evolution of the X-inactivation mechanism. In contrast, there is no obvious theoretical reason why the pseudoautosomal region should be deficient in expressed genes, although only two have described in man so far.

ACKNOWLEDGEMENTS

We would like to acknowledge the contributions made by P. Tippett and R. Sangar to the analysis of "*YG*". We are also grateful for the editorial assistance provided by C. Middlemiss.

REFERENCES

1. de la Chapelle, A: The 1985 human gene map and human gene mapping in 1985. Human Gene Mapping 8. *Cytogenet Cell Genet*, 40:1-7, 1985.

2. Darling, SM, Banting, GS, Pym, B, Wolfe, J, Goodfellow, PN: Cloning an expressed gene shared by the human sex chromosomes. Proc Natl Acad Sci USA, 83:135-139, 1986.

3. Darling, SM, Goodfellow, PJ, Pym, B, Banting, GS, Pritchard, C, Goodfellow, PN: Molecular Genetics of MIC2: a gene shared by the human X and Y chromosomes. Cold Spring Harbor Symp Quant Biol 51: in press, 1986.

4. Ohno, S: Sex chromosomes and sex-linked genes. New York: Springer Verlag. 1967

5. Lyon, MF: Evolution of X-chromosome inactivation in mammals. Nature, 250:651-653, 1974.

6. Charlesworth, B: A model for the evolution of Y chromosomes and dosage compensation. Proc Natl Acad Sci USA, 75:5618-5622, 1978.

7. Koller, PC, Darlington, CD: The genetical and mechanical properties of the sex chromosomes. 1. Rattus norvegicus. J Genet, 29:159-173, 1934.

8. Burgoyne, PS: Genetic homology and crossing over in the X and Y chromosomes of mammals. Hum Genet, 61:85-90, 1982.

9. Bengtsson, BO, Goodfellow, PN: The effects of recombination between the X and Y chromosomes. Ann Hum Genet, 1987, in press

10. Keitges, E, Rivest, M, Siniscalco, M, Gartler, SM: X-linkage of steroid sulphatase in the mouse is evidence for a functional Y-linked allele. Nature, 315:226-227, 1985.

11. Goodfellow, P, Pym, B, Mohandas, T, Shapiro, LJ: The cell surface antigen locus, MIC2X, escapes X-inactivation. Am J Hum Genet, 36:777-782, 1984.

12. de la Chapelle, A: The etiology of maleness in XX men. Hum Genet, 58:105-116, 1981.

13. Vergnaud, G, Page, DC, Simmler, MC, Brown, L, Rouyer, F, Noel, B, Botstein, D, de la Chapelle, A, Weissenbach, J: A deletion map of the human Y chromosome based on DNA hybridization. Am J Hum Genet, 38:109-124, 1986.

14. Goodfellow, P, Darling, S, Wolfe, J: The human Y chromosome. J Med Genet, 22:329-344, 1985.

15. Kratzer, PG, Gartler, SM: HPRT activity changes in pre-implantation mouse embryos. Nature, 274:503-504, 1978.

16. Davis, RM: Localisation of male determining factors in man: a thorough review of structural anomalies of the Y chromosome. J Med Genet, 18:161-195, 1981.

17. Daiger, SP, Chakraborty, R: Mapping the human Y chromosome. In: Sandberg AA (Ed) The Y chromosome. Alan R Liss Inc. New York 1985; 93-124

18. Simmler, MC, Rouyer, F, Vergnaud, G, Nysteom-Lahti, M, Ngo, KY, de la Chapelle, A, Weissenbach, J: Pseudoautosomal DNA sequences in the pairing region of the human sex chromosomes. Nature, 317: 692-697, 1985.

19. Cooke, HJ, Brown, WRA, Rappold, GA: Hypervariable telomeric sequences from the human sex chromosomes are pseudoautosomal. Nature, 317:687-692, 1985.

20. Rouyer, F, Simmler, MC, Johnsson, C, Vargnaud, G, Cooke, HJ, Weissenbach, J: A gradient of sex linkage in the pseudoautosomal region of the human sex chromosomes. Nature, 319:291-295, 1986.

21. Goodfellow, PJ, Darling, SM, Thomas, NS, Goodfellow, PN: A pseudo-autosomal gene in man. Science, 1986, in press

22. Goodfellow, PJ, Darling, SM, Pym, B, Pritchard, C, Goodfellow, PN: Homologies between the sex chromosomes of man. Proc 7th Int Hum Genet Congress 1987; in press

23. Renwick, JH: Progress in mapping human autosomes. Br Med Bull, 25:65-73, 1969.

24. Craig, IW, Tolley, E: Steroid sulphatase and the conservation of mammalian X chromosome. Trends Genet, 20:201-204, 1986.

25. Buckle, V, Fraser, N, Boyd, Y, Andria, G, Ballabio, A, Persico, G, Sebastio, G, Craig, I: Fine mapping of putative coding sequence for steroid sulphatase to Xp22.3: lack of Y-chromosomal copy. Proc 7th Int Congress Hum Genet 1987; in press

26. Goodfellow, P, Banting, G, Levy, R, Povey, S, McMichael, A: A human X-linked antigen defined by a monoclonal antibody. Somat Cell Genet, 6:777-787, 1980.

27. Goodfellow, P, Banting, G, Sheer, D, Ropers, HH, Caine, A, Ferguson-Smith, MA, Povey, S, Voss, R: Genetic evidence that a Y-linked gene in man is homologous to a gene on the X chromosome. Nature, 302:346-349, 1983.

28. Banting, GS, Pym, B, Goodfellow, PN: Biochemical analysis of an antigen produced by both human sex chromosomes. EMBO J, 4: 1967-1972, 1985.

29. Buckle, V, Mondello, C, Darling, S, Craig, IW, Goodfellow, PN: Homologous expressed genes in the human sex chromosome pairing region. Nature, 317:739-741, 1985.

30. Pritchard, C, Goodfellow, PN: Development of new methods in human gene mapping: selection for fragments of the human Y chromosome after chromosome-mediated gene transfer. EMBO J, 5:979-985, 1986.

31. Goodfellow, PN: Expression of the 12E7 antigen is controlled independently by genes on the human X and Y chromosomes. Differentiation, 23(S):35-39, 1983.

32. Goodfellow, PN, Tippett, P: A human polymorphism related to Xg blood groups. Nature, 289:404-405, 1981.

33. Shows, TB, et al: International system for human gene nomenclature. Cytogenet Cell Genet, 25:96-116, 1979.

34. Ferguson-Smith, MA, Sanger, R, Tippett, P, Aitken, DA, Boyd, E: A familial t (X:Y) translocation which assigns the Xg blood group to the region Xp22.3-pter. Cytogenet Cell Genet, 32: 272, 1982.

35. Race, RR, Sanger, R: Blood groups in man (6th edition). Blackwell Scientific Publications, Oxford, 1975

36. Tippett, P, Shaw, MA, Green, CA, Daniels, GL: The 12E7 red cell quantitative polymorphism: control by the Y-borne locus, Yg. Ann Hum Genet, 1986; in press

37. Jost, A, Vigier, B, Prepin, J, Perchellat, JP: Studies on sex differentiation in mammals. Recent Progr Horm Res, 29:1-41, 1973.

38. McLaren, A: Does the chromosomal sex of a mouse germ cell affect its development. In: Current problems in germ cell differentiation. British Soc Dev Biol Symp 7 (ed) McLaren A and Wylie CC. Cambridge University Press, 224-240

39. Chandra, HS: Sex determination: a hypothesis based on non-coding DNA. Proc Natl Acad Sci USA, 82:1165-1169, 1985.

40. Affara, NA, Florentin, L, Morrison, N, Kwok, K, Mitchell, M, Cook, A, Jamieson, D, Glasgow, L, Meredity, L, Boyd, E, Ferguson-Smith, MA: Regional assignment of Y-linked DNA probes by deletion mapping and thair homology with X-chromosome and autosomal sequences. Nucl Acids Res, 14:5353-5373, 1986.

41. Wachtel, SS: H-Y antigen and the biology of sex determination. Grune and Stratton, New York, 1983

42. Andrews, PW: The male specific antigen and sexual differentiation. In: Goodfellow P (ed) Genetic analysis of the cell surface. No.16. series B. Receptors and Recognition, London, Chapman Hall 1984; 159-160

43. McLaren, A, Simpson, E, Tomonari, K, Chandler, P, Hogg, H: Male sexual differentiation in mice lacking H-Y antigen. Nature, 312:552-555, 1984.

44. Townsend, ARM, Rothbard, J, Gotch, FM, Bahadur, G, Wraith, D, McMichael, AJ: The epitopes of influenza nucleoprotein recognised by cytotoxic T lymphocytes can be defined with short synthetic peptides. Cell, 44:959-968, 1986.

45. Bradley, MP, Heslop, BF: Elicitation of rapid and transient antibody response to H-Y antigen by intrasplenic immunisation. Transpl, 39:634-638, 1985.

46. Goulmy, E, Teomijtelen, A, Bradley, BA, Van Rood, JJ: Y-antigen killing by T cells of women is restricted by HLA. Nature,

266:544-545, 1979.

47. Burgoyne, PS, Levy, ER, McLaren, A: Spermatogenic failure in male mice lacking H-Y antigen. Nature, 320:170-172, 1986.

48. Dronamaraju, KR: Y-linkage in man. Nature, 201:424, 1964.

49. Tiepolo, L, Zuffardi, O: Localisation of factors controlling spermatogenesis in the non-fluorescent portion of the human Y chromosome long arm. Hum Genet, 34:119-124, 1976.

50. Alvesalo, L, de la Chapelle, A: Tooth sizes in two males with deletions of the long arm of the Y-chromosome. Ann Hum Genet, 45:49-54, 1981.

51. Koenig, M, Moison, JP, Heilig, R, Andre, G, Mandel, J-L: Homologies between the X and Y chromosomes analysed with DNA probes. Human Gene Mapping 8. Cytogenet Cell Genet, 40:670-671, 1985.

52. Freytag, SO, Bock, GHO, Beaudet, AC, O'Brien, WE: Molecular structures of the human argininosuccinate synthetase pseudogenes. J Biol Chem, 259:3160-3166, 1984.

THE DETECTION OF Y DNA IN INTERSEX SUBJECTS WITH MOSAICISM FOR A Y CELL LINE

Paul G. McDonough, Sandra P. Tho, John J. Trill,
J. Rogers Byrd, and Jay A. Tischfield

Departments of Obstetrics and Gynecology, Endocrinology and Anatomy, and the Human Genetics Institute, Medical College of Georgia, Augusta, Georgia 30912

ABSTRACT

Repetitive copy DNA sequences corresponding to Yq12 were detectable in mixtures of 2.5ng of male DNA and 997.5ng of 45,X female DNA. Hybridization with the same, repetitive copy Y DNA probe (pS4) was performed on the DNA of 20 intersex subjects with a normal or variant Y chromosome. DNA from the same subjects was hybridized to a single copy probe (4B-2) mapped to the Yq11 region. DNA from subjects in category A (n = 8), with cytologically normal Y chromosomes, hybridized to both deoxyribonucleic acid probes. DNA from subjects in category B (n = 2), which included a variant Y chromosome which was negative for Q-banding but positive for C-bands, hybridized with the distal pS4 and proximal 4B-2 probes. DNA from subjects in category C (n = 10), with variant Y chromosomes which were uniformly negative for Q- and C-bands, did not hybridize with the pS4 probe. DNA from three of the ten subjects in category C did hybridize to the more proximal sequence-detecting 4B-2 probe. DNA from the remaining seven subjects in category C did not hybridize with either of the DNA probes.

INTRODUCTION

Techniques for the analysis of DNA may be useful for detection of DNA from the Y chromosome in putative mosaic 45,X/46XY subjects or for confirmation of the presence of a suspected Y fragment, to assist in the identification of subjects at risk for dysgenetic tumors. The utility of molecular probes for the Y chromosome in this clinical situation will depend upon their potential to detect: 1) small amounts of Y chromosomal DNA in mosaic 45,X/46,XY subjects, and 2) structurally altered Y chromosomes, especially del(Y)(q12). Disteche et al. recently constructed a Y-specific DNA probe from an 850 base pair sequence of nucleotides derived from the 2.1 Kb, *Hae* III, repetitive DNA fragment visible on gels stained with ethidium bromide[1,2]. After digestion of DNA with *Mbo* I, hybridization by Southern blots to this repetitive Y probe (pS4) allows a clear difference to be discerned between male and female DNA, with male DNA exhibiting a sex-specific band that corresponds to a fragment of 2.3 Kb. Recent evidence suggests that the upper limits of the pS4 sequences on the Y long arm are defined by the upper limits of C-banding and hence, they are contained within Yq123.

Burk et al[4] isolated a single copy DNA probe, 4B-2, which maps to the proximal portion of the long arm of the human Y chromosome (subcentromeric -q11), placing it entirely in the euchromatic region.

Three different but related experiments, using a combination of these two different Y DNA probes were designed to evaluate the sensitivity and specificity of molecular techniques for the detection of Y DNA in mosaic 45,X/46,XY intersex subjects[5].

MATERIALS AND METHODS

General

Isolation of Plasmids

The DNA probe pS4 (supplied by D. Hoar[9]) which hybridizes to repetitive sequences on the Y chromosome, consists of a 850 base pair, Pst I, nucleotide sequence derived from the 2.1 Kb, Hae III fragment of DNA regionalized to the heterochromatic region of the long arm of the Y chromosome[1]. The probe 4B-2 (supplied by K. Smith[4]) is a recombinant phage construct, containing a single copy 3.3 Kb Eco RI fragment, which maps to the proximal portion of the euchromatic region of the long arm of the Y chromosome. The recombinant plasmid was constructed by Eco RI digestion of the recombinant phage 4B-2, and insertion of the gel-purified fragment into the Eco RI site of pBr 3224. Both plasmids were propagated in *E. Coli* HB101. The bacteria were grown in Luria broth which contained ampicillin. The plasmids were amplified by administration of chloramphenicol and isolated by the method of Ish-Horowitz[6]. The plasmids were digested with restriction enzymes and the resultant fragments were purified by elecrophoresis on a 0.8% low melting point agarose gel followed by purification with a NACS Prepak column (BRL), according to the manufacturer's recommendations.

Isolation of DNA

Total genomic DNA was isolated from 20-30 ml of EDTA-whole blood. The blood was mixed 1:1 with lysis buffer (1mM MgCl, 1mM NaH_3PO_4, pH 6.5, 0.8% NP-40 and 0.4% deoxycholate) and centrifuged in a JA20 rotor at 10,000 rpm for 30 min at 4°C. The pellet containing nuclei was dissolved in 10 ml of TEN (10mM Tris, pH 8.3, 150 mM NaCl, 5 mM EDTA); 3.5 ml of 5 M $NaClO_4$; 1.5 ml of 10% SDS and digested for 1 hour at 65°C with proteinase K. The DNA was extracted twice with chloroform: isoamyl alcohol (24:1) and precipitated by the addition of 1/2 volume of 7.5M ammonium acetate and 2 volumes of absolute alcohol. The DNA was lyophilyzed and subsequently dissolved in 1X TE plus 1X SSC, digested with pancreatic RNase at 37°C for 30 min. and redigested at 65°C for 1 hour with proteinase K. The DNA was then reextracted with chloroform:isoamyl alcohol and precipitated with etha l.

Digestion by Restriction Enzymes, Gel Electrophoresis and Hybridization

Genomic DNA, after extraction, was digested with either *Hae* III, *Mbo* I, *Eco* RI or *Hind* III under the conditions recommended by the supplier (BRL). The DNA was electrophoresed on 0.8% agarose gels and blotted, by the method of Southern[7] to Gene Screen Plus filters (New England Nuclear). The filters were hybridized to either the pS4 or 4B-2 probes which were ^{32}P-labeled by nick translation[8], and washed as recommended by the supplier. After the final wash, the filters were washed for an additional 90 min. at 65°C in 0.1X SSC with 1% SDS.

Specific Experiments

Experiment I: Dilutional Gel of Male and Female 45,X DNA, Digested with Hae III, Mbo I, and Hybridization with pS4 Plasmid

Decreasing amounts of normal male DNA (100 ng to 10 pg) were mixed with female DNA from a 45,X gonadal dysgenesis subject, to give a final concentration of 1 g DNA per lane, and digested separately with either *Hae* III or *Mbo* I. 1 g of each of the digested samples of DNA was electrophoresed on a 0.8% agarose gel, blotted and hybridized to a ^{32}P-labeled 850 bp *Pst* I fragment isolated from pS4.

Experiment II: Digestion with Mbo I and Hybridization with pS4 DNA, of DNA from Study Subjects

This experiment was designed to compare the specificity of molecular and cytogenetic techniques in the assessment of normal and altered Y chromosomes. Genomic DNA from each subject (n = 20) and their fathers (n = 14) was digested with *Mbo* I, and 1.5 g DNA/well was electrophoresed on a 0.8% agarose gel, blotted, and hybridized to the pS4 probe, as above.

Clinical Subjects

All study subjects (n = 20), with one exception, were 45,X/46,XY mosaics with normal Y (n=8) or variant Y (n=12) chromosomes. (see Table I) In one of the 12 subjects with a variant Y chromosome, a 45,X cell line was not detected (subject 9). The single cell line from this subject with an altered Y chromosome was obtained from a maternal age-indicated culture of amniotic fluid at 16 weeks of gestation. The product of this pregnancy, now three months of age, is a normal-appearing male with bilateral scrotal testes. All other karyotypes (n=19) were derived from analysis of cells from peripheral blood.

The clinical phenotypes of the 20 subjects with normal and variant Y chromosomes ranged from phenotypic males with bilateral scrotal testes (n = 3) to females with Turner or quasi-Turner phenotypes (n = 17). The 17 phenotypic females had bilateral streak gonads (n = 12) or a unilateral streak gonad with a contralateral testis (n = 5). Dysgenetic tumors occurred in three of the subjects. In two of the 17 phenotypic females, one or both rudimentary gonads had been replaced by a tumor. One patient (aged 20 years) had a unilateral gonadoblastoma (subject 17); and the other, at 15 years of age, had had bilateral tumors, dysgerminoma and gonadoblastoma (subject 1). One enigmatic subject (No. 19) with a left

Table I. Tabulated Clinical, Cytogenetic, and DNA Data on 20 Study Subjects.

Subjects in category A (1 through 8) have cytologically normal Y chromosomes and their DNA hybridizes with both pS4 and 4B-2 probes. Category B subjects (9, 10) represent one subject with loss of Q-banding, but intact C-bands, and one subject with a dicentric Y chromosome in multiple cell lines. Subject 9 was positive for C-heterochromatin and his DNA still hybridized with both probes. Subjects in category C, with del(Y)(ql2), fall into two categories depending on whether or not their DNA hybridizes with the proximal 4B-2 probe. DNA from subjects in the C1 category (11 - 13) hybridizes with the 4B-2 sequences, whereas the DNA from the remaining subjects (14 -20) does not demonstrate a 4B-2 band. (C2)

Case	Age*	Phenotype	Karyotype % 46,XY Cells Counted	Morphology of Y	Banding of Y: C	Banding of Y: QF	Banding of Y: G	pS4 Mbo I	4B-2 Eco RI	4B-2 Hind III	
1	40	♀ with RSG†	10% (50)	n	+	+	+	+	+	+	
2	25	♀ with RSG	16% (50)	n	+	+	+	+	+	+	
3	20	♀ with RSG	20% (50)	n	+	+	+	+	+	+	
4	15	♀ with RSG	88% (50)	n	+	+	+	+	+	+	
5	21	♀ with RSG	82% (50)	n	+	+	+	+	+	+	
6	3	♀ with unilateral labial testis	84% (50)	n	+	+	+	+	+	+	
7	3	♂ with bilateral scrotal testes	87% (100)	n	+	+	+	+	+	+	
8	25	♀ with unilateral abd testis	94% (50)	n	+	+	+	+	+	+	
9	3mos	♂ with bilateral scrotal testes	100% (50)	del(Y)(q12)	+	–	–	+	+	+	
10	15	♀ with unilateral abd testis	94% (50)	dic(Y)(qter→p11::p11→qter)	+	+	+	+	+	+	
11	25	♀ with RSG	14% (50)	del(Y)(q12)	–	–	–	–	+	+	C_1
12	7	♀ with RSG	64% (50)	del(Y)(q12)	–	–	–	–	+	+	C_1
13	8mos	♀ with RSG	40% (50)	del(Y?) acentric fragment?	–	–	–	–	+	+	C_1
14	30	♀ with RSG	52% (50)	r(Y?)(p11q12)	?	–	–	–	–	–	C_2
15	7	♀ with unilateral labial testis	50% (50)	del(Y)(q12)	–	–	–	–	–	–	C_2
16	28	♀ with RSG	18% (50)	del(Y?) acentric fragment?	–	–	–	–	–	–	C_2
17	40	♀ with RSG†	68% (50)	del(Y?) acentric fragment ?	–	–	–	–	–	–	C_2
18	15	♀ with RSG	92% (50)	del(Y)(q12)	–	–	–	–	–	–	C_2
19	15	♀ with bilateral* testes and unilateral streak	76% (50)	del(Y)(q12) var(22)(p13,QFQ 4,5)	–	–	–	–	–	–	C_2
20	30	♂ azoospermia with bilateral scrotal testes	72% (50)	del(Y)(q12)	–	–	–	–	–	–	C_2

*: At the time of this study; †: Dysgenetic tumors; Case 1: Dysgerminoma; Case 17: Gonadoblastoma; Case 19: Embryonal carcinoma; RSG: Rudimentary streak gonads

(Reproduced, with permission, from Am J Obstet Gynecol 154:737-748, 1986.)

+ = Dysgenetic Tumor
RSG = rudimentary streak gonads
C = heterochromatin
QF = Q-banding
G = Giemsa banding

intra-abdominal testis and right contralateral streak, both of which were extirpated at 5 years of age, developed a tumor in a second right inguinal testis at 15 years of age. The clinical and cytogenetic data are tabulated in Table I.

Cytogenetic Analysis

Prior to analysis of their DNA, karyotypes of all subjects were reanalysed from cells in peripheral blood to update the percentage of mosaicism, to reclarify the structure of variant Y chromosomes, and to ensure that the molecular analysis was performed on the same sample from which the current percentage of mosaicism had been determined. The karotype of the father of each proband, when available (n = 14) was also obtained for comparison of Y structure. A minimun of 50 cells was counted, and 5 cells were karyotyped in all study samples. G, Q, and C banding were performed on every study sample prior to analysis of DNA in the same sample.

Based on the cytogenetic findings, the study subjects were placed into 3 categories at this time. Subjects in category A had normal Y chromosomes by all cytologic parameters. Of the subjects in category B, one had a dicentric Yq, and the other had a deleted Yq which eliminated Q-banding but preserved C-bands. Subjects in category C were cytogenetically uniform, in that they all appeared to have deletions of Yq which eliminated both Q and C bands (Table I).

Experiment III: Digestion by EcoRI and Hind III, and Hybridization with 4B-2 DNA, of DNA from Study Subjects

To categorize further the deleted Y chromosomes (negative Q, negative C) in subjects in category C, a third experiment was performed. Using normal males, females, and subjects from category A as controls, all individuals in category C (n = 10), were further analyzed with the single copy DNA probe 4B-2. This probe, provisionally localized to the paracentric region of Yq, is proximal to the pS4 sequences[4].

DNA from subjects in category A and category C was digested with either *Eco* RI or *Hind* III. The DNA was electrophoresed (7.5 g/well) on 0.8% agarose gels, blotted and hybridized to a ^{32}P-labeled 3.3 Kb *Eco* RI insert, isolated from the recombinant phage 4B-2, and subcloned into the *Eco* RI site of pBR 322.

RESULTS

Experiment I: Dilutional Gel of Male and Female 45 X DNA Digested with Hae III, Mbo I, and Hybridization with pS4 Plasmid

Normal male and female DNA, restricted with *Hae* III and probed with pS4, exhibited a 2.1 Kb band of hybridization. This single band, though less intense in female DNA, was present in both sexes. In DNA from controls and study subjects, a distinct male-specific band was not evident with the *Hae* III digest and pS4 probe.

Normal male and 45,X female DNA, restricted with Mbo I and probed with pS4, showed clear differences[9]. Male DNA, exhibiting 10-fold more hybridization than female DNA, generated a 2.3 Kb male-specific band. In gels containing 900 ng of 45,X DNA and 100 ng of male DNA, this Y-specific 2.3 Kb band was easily detected. This band was visible until the amount of male DNA was less than 2.5 ng[5]. This result represents a detection of 1 part male DNA in 400 parts 45,X DNA (Figure 1).

Experiment II: Digestion by MboI and hybridization with pS4 DNA, of DNA from Study Subjects

Clinical Subjects and Cytogenetic Analysis

The clinical and cytogenetic data from the 20 study subjects are tabulated in Table I.

Category A (n = 8)

Subjects 1 through 8, assigned to category A, and their fathers, had morphologically normal Y chromosomes which were positive for Q, G, and C bands. The percentage of 46,XY cells in peripheral blood ranged from a low value of 10% (subject 1) to 94% (subject 8). In this category, there is one subject with bilateral dysgenetic tumors (subject 1), one

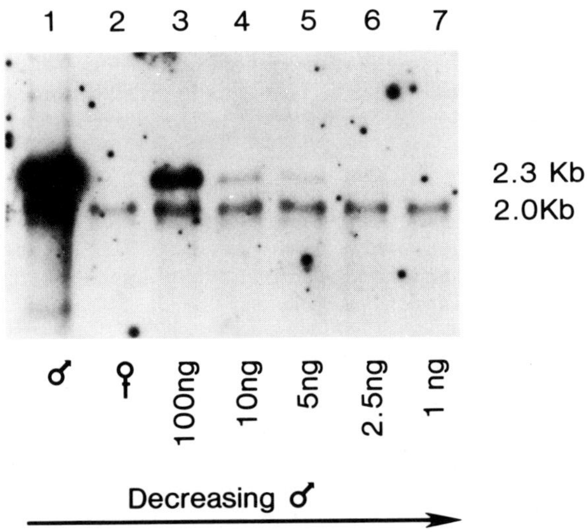

Figure 1: A dilutional gel of *Mbo* I digested DNA, probed with pS4. Lanes 1 and 2 represent normal male and 45,X female control samples. Lane 3 represents a 1 μg mixture of 900 ng 45,X DNA and 100 ng normal male DNA. Lanes 4-7 represent decreasing amounts of male DNA mixed with proportionately increasing amounts of 45,X DNA, with 1 μg total DNA in each lane. The Y-specific 2.3 Kb band is evident when 5 ng male DNA was present, and possibly also when only 2.5 ng male DNA was present. The 2.5 ng level represents a mixture of 2.5 ng/997.5 ng or 1 part male in 400 parts 45,X DNA. (Reproduced, with permission, from *Am J Obstet Gynecol* 154:737-748, 1986.)

subject with bilateral scrotal testes (subject 7), and there are two subjects with unilateral testis and contralateral streak gonad (subjects 6, 8).

Category B (n = 2)

One subject (9) is placed in this category because of his cytogenetic uniqueness. The morphology of the Y chromosome is intermediate between the normal Y (category A), and the morphologic variants seen in category C. The subject is a normal male in terms of phenotype, as initially ascertained in amniocyte culture, in whom a 45,X cell line has not been detected. The Y chromosome in the single cell line is completely negative for distal long arm Q banding, but a distinct terminal heterochromatic band is evident with C-banding[3].

The second subject in Category B (subject 10) has a dicentric Y chromosome, with loss of Yp and duplication of Yq. This subject is the only one in our study who is theoretically disomic for DNA on the long arm of the Y chromosome.

The fathers of both subjects (9, 10) in category B had cytologically normal Y chromosomes with distinct Q and C-bands.

Category C (n = 10)

Category C is represented by 10 subjects with clear morphologic deletions in their Y chromosomes. All subjects in category C are negative for Q and C-bands. The size of the Y chromosome is reduced in all subjects, and in four subjects it is only a minute centric fragment, an acentric fragment or a ring (subjects 13,14,16,17). No further attempts, utilizing cytogenetic techniques, were made to resolve the precise nature of these deleted chromosomes. The percentage of the variant Y cell line in subjects in category C ranged from 14% (subject 11) to 92% (subject 18). The paternal Y chromosomes of subjects in category C were morphologically normal and positive for Q and C-bands, in those cases where the paternal chromosomes were examined.

In this category there are two subjects with tumors (subjects 17 and 19), one azoospermic male with testicular failure and small bilateral scrotal testes (subject 20), and two subjects with unilateral testis and contralateral streak gonad (subjects 15 and 19).

DNA Analysis (Mbo I, pS4)

The presence or absence of the 2.3 Kb, male-specific band is indicated for each individual subject in Table I. The autoradiographs, representative of each of the three cytogenetic categories, are seen in Figure 2.

All subjects in category A, with morphologically normal Y chromosomes, gave a distinct 2.3 Kb band. This band was present in every subject (1-8) in category A, even at the lowest level of XY mosaicism (10%) (Figure 2, Lane 3).

The one subject in category B, (subject 9) negative for Q, but positive for C-bands, generated the 2.3 Kb male-specific band, but at a reduced intensity (Figure 2, Lane 4)[3]. The one subject with dicentric

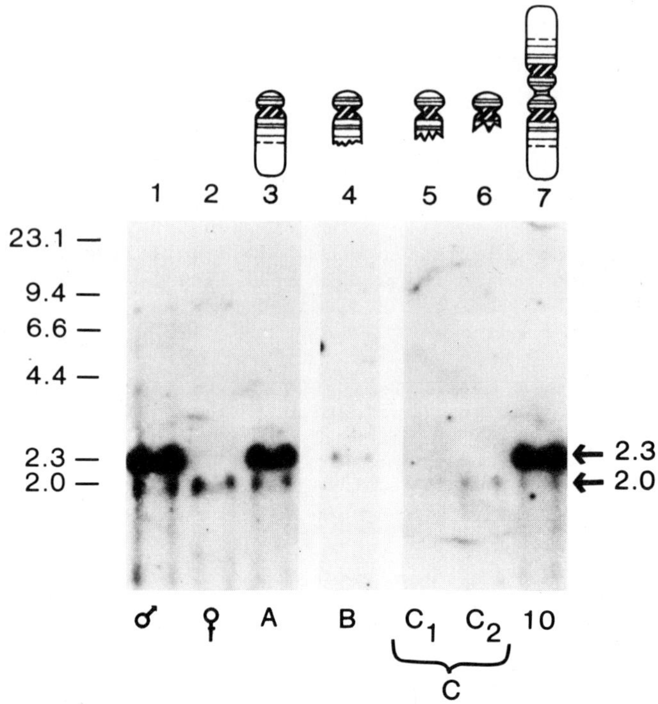

Figure 2: An autoradiogram of total genomic DNA digested with Mbo I and probed with pS4. The lanes from left to right are: Lane (1), 1.5 μg of control normal male DNA; Lane (2), 1.5 μg of control normal female DNA; Lane (3), 1.5 μg of DNA from case #1, representative of cases 1-8 in the normal Y category (category A); Lane (4), 600 ng (0.6 μg)of DNA from case #9, deleted of Q-bands, but still positive for C-banding (category B); Lanes (5) and (6), 1.5 μg of DNA from cases #11 and 19, representative of subjects with deleted Q and C-bands (category C); Lane (7) 1.5 μg DNA from subject #10 with dicentric Y long arm. (Reproduced, with permission, from Am J Obstet Gynecol 154:737-748, 1986.)

Yq (subject 10) exhibits the 2.3 Kb band in increased intensity, as compared to a normal male control (Figure 2, Lane 7).

No subject in category C (11-20) devoid of Q and C-bands, demonstrated the 2.3 Kb band (Figure 2, Lanes 5,6).

Experiment III: Digestion by Eco RI and Hind III, and Hybridization with 4B-2 DNA, of DNA from Study Subjects

Analysis of DNA

Digestion of normal male DNA with Hind III and blot hybridization with 4B-2 yields hybridization bands of 2.3 Kb and 1.8 Kb. These bands are not seen in autoradiograms of digested and blotted female DNA (Figure 3).

Digestion of genomic DNA with Eco RI, and blot hybridization with the 4B-2 probe yields a single, male-specific hybridization band of 3.3 Kb. Normal female DNA does not generate this band (Figure 4).

The results of probing with 4B-2, after restriction with Hind III and Eco RI, are tabulated for individual subjects in Table I, and representative autoradiograms are presented in Figures 3 and 4.

Normal males and all subjects in category A demonstrated the male-specific 2.3 Kb, 1.8 Kb Hind III and 3.3 Kb Eco RI fragments. Three of the 10 subjects in category C, (11,12,13) whose DNA failed to hybridize with the pS4 probe, hybridized with the 4B-2 single copy probe after restriction with both Hind III and Eco RI. Representative autoradiograms from this group, now referred to as category C_1 are depicted in Figures 3 and 4 (Lanes 4). The remaining 7 subjects in category C (14-20) had DNA which failed to hybridize with the 4B-2 probe and these subjects are, hence, assigned to category C_2 (Figures 3 and 4, Lanes 5).

DISCUSSION

This series of three experiments addresses the ability of specific molecular techniques to identify human Y chromosomal DNA. The clinical usefulness of these techniques will depend upon their sensitivity and precision. The male-specific repetitive DNA probe pS4, in mixtures of 45,X and 46,XY DNA, was able to detect as little as 2.5 ng of male DNA. This probe detected male-specific DNA sequences in all subjects, except those with Yq deletions that involve loss of both Q and C bands. The DNAs from the 10 individuals with Y chromosome deletions proximal to the termination of C-banding (category C) did not hybridize with the pS4 probe. The one subject (#9) with a Yq deletion, who demonstrated a distinct 2.3 Kb Y-specific band, still had a small, distinct C-band on the Y long arm (category B). The presence of the Mbo I, pS4 2.3 Kb band in this subject, who was devoid of Q-bands but positive for C-banding, suggests that the sequences which correspond to the pS4 probe are defined on Yq by the upper limits of C-banding, and are not restricted to the Q-positive region. The molecular data, contrary to general cytogenetic opinion, suggest that the chromosomal regions that represent the C and Q-bands do not overlap in their entirety. C-heterochromatin, as defined by the upper limits of the pS4 sequences, seems to extend more proximally than Q banding[3].

Experiment II clearly defines the limits of the pS4 sequences in this study group. A 2.3 Kb male-specific band is not generated by the DNA from any of the 10 subjects with Yq deletions who have lost both C and Q-bands (category C). One may assume that these 10 individuals have a Y deletion that extends up into the euchromatic portion of the long arm of Y.

From the results of Experiment III, it is apparent that even these 10 individuals in category C, who appear to be cytogenetically indistinguishable, are not homogeneous in terms of DNA analysis. The DNA from 3 of these subjects, when restricted with Eco Rl and Hind III, hybridizes with the more proximally located single copy 4B-2 (category C_1). The DNA from the remaining 7 subjects, with presumably more extensive deletions of the Y chromosome, does not demonstrate the

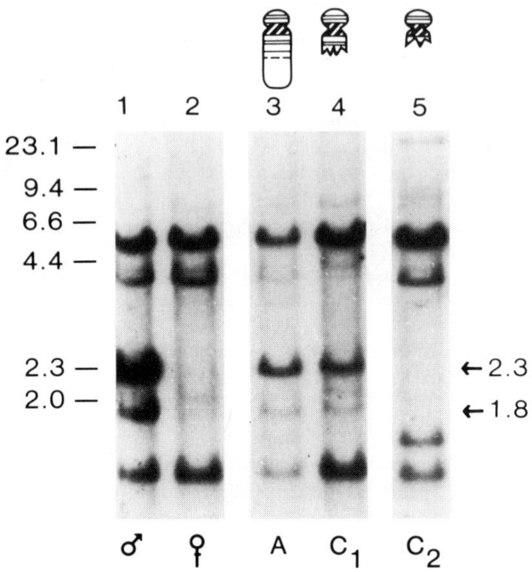

Figure 3: Autoradiograms of total genomic DNA digested with <u>Hind</u> III, followed by hybridization with 4B-2. Lanes 1 and 2 represent normal male and normal female DNA. Normal male DNA, after restriction with <u>Hind</u> III, has 2.3 Kb and 1.8 Kb fragments not present in female DNA. Lane 3 represents one subject from Category A, with a morphologically normal Y chromosome. Lane 4 is representative of subjects (11 through 13) from Category C in whom the Q and C-bands have been deleted, and whose DNA hybridizes with the 4B-2 probe and demonstrates the two male-specific fragments (Category C_1). Lane 5 is representative of the remaining 7 subjects (14 through 20) in Category C, whose DNA does not hybridize with the 4B-2 probe (Category C_2). (Reproduced, with permission, from <u>Am J Obstet Gynecol</u> 154:737-748, 1986.)

male-specific bands with this single copy probe (category C_2). This result suggests that the largest DNA deletions are present in these 7 individuals, who are more or less cytogenetically indistinct from the other 3 del(Y) subjects (Figure 5). Alternatively, some of these 7 subjects may have a complete absence of Yq DNA. The severely variant chromosomes in these subjects may be isochromosomes for Yp, or even represent fragments of X.

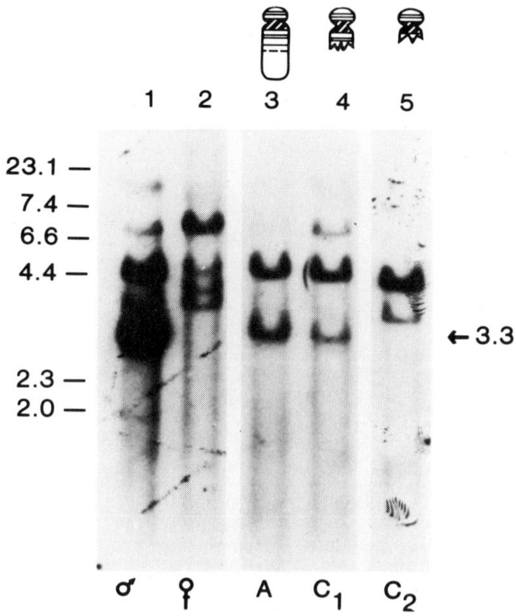

Figure 4: Autoradiograms of total genomic DNA, digested with Eco RI and then by hybridized with 4B-2. Male DNA (Lane 1) has a 3.3 Kb band not present in female DNA (Lane 2). Subjects in category A and C_1 generate this band, (Lanes 3 and 4) but it is not seen for subjects in category C_2 subjects (Lane 5). (Reproduced, with permission, from Am J Obstet Gynecol 154: 737-748, 1986.)

It is apparent from these three experiments that cytogenetic techniques will continue to be necessary to define the gross morphology of variant Y chromosomes. Analysis of DNA can then be selectively used to characterize more precisely the limits of these alterations (Figure 5).

Characterizing the DNA composition of the variant or altered Y chromosome may provide insight into the critical determinants of oncogenesis and gonadal differentiation. Sensitive and precise techniques to detect covert Y chromosomal DNA, and those which are capable of identifying variant Y chromosome DNA, will help to define the population of individuals at risk for dysgenetic tumors, and may provide clinical insight into their variant gonadal development. The widespread application of DNA technology to detect all variant Y chromosomes will require centromeric-specific Y probes[10].

ACKNOWLEDGEMENTS

We gratefully acknowledge Kirby Smith and David Hoar for their gifts of the pS4 and 4B-2 DNA probes.

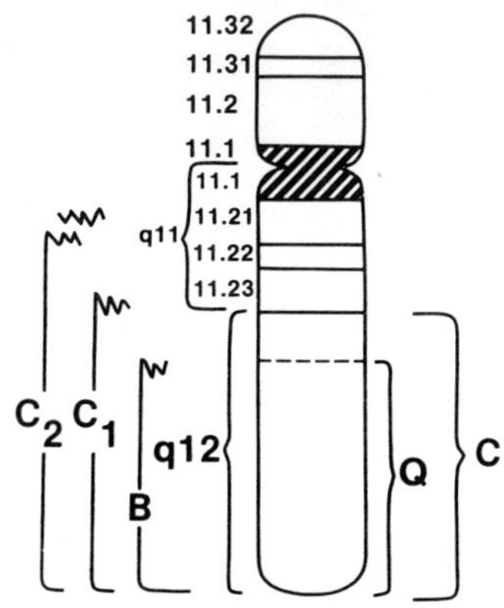

Figure 5: A schematic representation of the extent of the deleted Yq in each of the three categories of Y variants is compared to the normal category A. In variant B, the breakpoint extends to the top of the Q-band, but does not reach to the top of the C-band. DNA from a type B variant, similar to a cytogenetically normal Y (category A), hybridizes with both the pS4 and the 4B-2 probes. The type C_1 variant is indistinguishable cytogenetically from the type C2 variant. Both variants are devoid of Q- and C-bands. DNA from type C_1 does not hybridize with pS4, but does hybridize with 4B-2, indicating a more extensive deletion than is present in type B. DNA from varient type C_2 does not hybridize with either probe and the subjects are presumed to have the most extensive deletion into the proximal portion of the long arm of Y, or alternatively to have isochromosomes for Yp with complete absence of long arm DNA. (Reproduced, with permission, from Am J Obstet Gynecol 154:737-748, 1986.)

REFERENCES

1. Disteche, C, Luthy, D, Haslam, DB, Hoar, D: Prenatal identification of a deleted Y chromosome by cytogenetics and a Y-specific repetitive DNA probe. Human Genetics, 67:222-224, 1984.

2. Cooke, H: Repeated sequences specific to human males. Nature, 262:182-186, 1976.

3. McDonough, PG, Tho, SPT, Trill, JJ, Byrd, JR, Hoar, DI, Smith, KD, Tischfield, JA: Upper limits of pS4 sequences on Y long arm mapped by differential Q- and C-banding. Human Gene Mapping 8, Helsinki Conference (1985); Eighth International Workshop on Human Gene Mapping Cytogenetics and Cell Genetics, 40(1-4):696, 1985.

4. Burk, RD, Patrick, MA, Smith, KD: Characterization and evaluation of a single copy sequence from the human Y chromosome. Mol and Cell Biology, 5:576-581, 1985.

5. McDonough, PG, Tho, SP, Trill, JJ, Byrd, JR, Reindollar, RH, Tischfield, JA: Use of two different DNA probes to detect Y chromosome DNA in subjects with normal and altered Y chromosomes. Am J Obstet Gynecol, 154:737-748, 1986.

6. Ish-Horowitz, Maniatis, T, Fritsch, EF, Sambrook, J: Molecular Cloning: A Laboratory Manual. Cold Spring Harbor Laboratory, Cold Spring Harbor, NY 90-91, 1985.

7. Southern, E: Detection of specific sequences among DNA fragments separated by gel electrophoresis. J Mol Biol, 98:503-517, 1975.

8. Stambrook, PJ, Dush, MK, Trill, JJ, Tischfield, JA: Cloning of a functional human adenine phosphoribosyltransferase (APRT) gene: Identification of a restriction fragment length polymorphism and preliminary analysis of DNAs from APRT-deficient families and cell mutants. Somat Cell Mol Genet, 10:359-367, 1984.

9. Hoar, DI, Haslam, DB, Starozik, DM: Improved direct molecular diagnosis and rapid fetal sexing. Prenatal Diagnosis, 4:241-247, 1984.

10. Wolfe, J, Darling, SM, Erickson, RP, Craig, IW, Buckle, VJ, Rigby, PWJ, Willard, HF, Goodfellow, PN: Isolation and characterization of an alphoid centromeric repeat family from the human Y chromosome. J Mol Biol, 182:477-485, 1985.

EVIDENCE THAT THE X-Y PAIRING/RECOMBINATION REGION MAY BE INVOLVED IN *tda-1* INHERITED SEX REVERSAL

Claude M. Nagamine* and Gloria C. Koo**

*Memorial Sloan-Kettering Cancer Center
New York, New York
**Merck, Sharp and Dohme Research Laboratories
Rahway, NJ 07065

INTRODUCTION

In mammals, the Y chromosome controls differentiation of the testes[1]. The testes, once differentiated, secrete Mullerian inhibiting substance and testosterone, both of which impose a male phenotype on a fetus inherently destined to be female[2]. Indeed, even primordial germ cells in the normal, male fetus autodifferentiate into oocytes unless situated within the seminiferous tubules of the testis[3-5].

When the Y chromosome of the longtailed European house mouse, *Mus musculus domesticus*[6-8], is placed into the C57BL/6J (B6) inbred mouse genome, XY fetuses develop into hermaphrodites, i.e. they possess both ovarian and testicular tissues, or they develop into females[9-12]. Karyotype analyses have excluded chromosomal mosaicism and abnormalities of the *M. m. domesticus* Y chromosome (Y^{Dom}) as possible causes for this phenomenon[10]. Apparently, the Y^{Dom} chromosome is incapable of inducing complete masculinization in B6.

One interpretation of these results is that the testis-determining gene on the Y^{Dom} chromosome (*Tdy*^{Dom}) interacts improperly with a putative B6 recessive gene named testis-determining, autosomal-1 (*tda-1*). All mice possessing the Y^{Dom} chromosome, which are also homozygous for *tda-1*^{B6}, develop either as hermaphrodites or as XY females[9-11]. *tda-1* has been tentatively assigned to an autosome or to the X-Y pairing-recombination region, but is known not to be present in the centromeric regions of chromosomes 1,3,5,8,9 and 11-17[11].

In sex reversal mediated by *tda-1*, despite the fact that, during fetal development, XY fetuses homozygous for *tda-1*^{B6} never differentiate a normal testis[10,13], the phenotypes of XY adults range from male phenotypes with bilateral testicular gonads and overt hermaphrodites with testicular and ovarian gonads, to female phenotypes with bilateral ovarian gonads. At the gross anatomical level, the testicular gonad can be indistinguishable from a normal testis, and even at the histological level, a single, developing oocyte may be the only indication of hermaphroditism[13]. At present, nothing is known about why the XY gonad differentiates into an ovotestis which is almost indistinguishable from a normal testis in one fetus but into an ovary in another. It has been suggested that other genes in the B6 genome must be homozygous in order to inhibit completely the normal differentiation of testes in about half of the XY fetuses which are homozygous for tda-1^{B6} [11].

B6.Y^{Dom} overt hermaphrodites with normal male genitalia can be fertile (ref. 10, and our observations). According to Eicher's hypothesis[9-11], overt hermaphrodites should be homozygous for the tda-1^{B6} allele. If other genes in the B6 genome must be homozygous for formation of XY females, then one would expect an increase in female phenotypes when B6.Y^{Dom} overt hermaphrodites are backcrossed to B6. In this report, we provide data on the post-natal progeny of B6.Y^{Dom} overt hermaphrodites and on fetal progeny of B6.Y^{Dom} crosses, using the steroid sulphatase (STS) locus as a marker. The data suggest that a gene or genes in the X-Y pairing/recombination region, centromeric to the steroid sulphatase locus, may be involved in generation of XY females.

METHODS

Mouse Strains

M. m. domesticus (Tirano, Italy) were obtained from Dr. Michael Potter. C3H/An mice were obtained from the Core Breeding Colonies maintained by Dr. E.A. Boyse at the Memorial Sloan-Kettering Cancer Center, NY. C57BL/6 mice were obtained from both the Jackson Laboratories (B6/J) and the Core Breeding Colonies (B6/Boy).

Generation of the B6.Y^{Dom}/Na Strain

M. m. domesticus males were mated with B5/J females and the F1 males backcrossed to B6/J[10]. The resulting strain has been designated B6.Y^{Dom}/Na[13,14]. These B6.Y^{Dom} male phenotypes of the N8 backcross generation, which should be consomic with B6/J[5], when mated to B6/Boy, resulted in hermaphroditism identical to that seen when they were mated to B6/J. We conclude that genetic drift between the two populations has not affected tda-1 sex reversal. Data obtained from both populations were pooled and designated as B6 unless noted otherwise.

Identification and Mating of B6.Y^{Dom} Overt Hermaphrodites

Six overt hermaphrodites of the N5 generation, possessing normal male genitalia, were identified under methoxyflurane anesthesia (Metofane, Pitman-Moore, Inc., NJ) at 6-8 wks of age. Internally, all of them possessed an ovarian gonad and a contralateral testicular gonad. After removal of the ovarian gonad, the hermaphrodites were allowed to recover, then backcrossed to B6.

Categorization of B6.Y^{Dom} Fetuses and Post-natal Progeny

Fetuses of known gestational age were obtained by examination of mating females for copulation plugs (day of plug = day 0 of gestation) or by comparison of fetuses to B6 fetuses of known gestational age[13]. Fetuses were categorized into male, female, or hermaphroditic phenotypes by the method of Whitten et al.[16]. Male phenotypes possessed bilateral testes, characterized by complete *tunicae albugineae* and well-developed seminiferous tubules. Female phenotypes possessed bilateral ovaries, characterized by a fine-grained appearance and absence of both tunicae albugineae and seminiferous tubules. In hermaphroditic phenotypes, one or both gonads had the tunica albuginea and seminiferous tubules present in only part of the gonad, the remainder of the gonad possessing ovarian characteristics.

Post-natal mice were categorized into male phenotypes, female phenotypes or overt hermaphrodites, according to the gross anatomy of the gonads and accessory ducts as seen under the dissecting microscope. Male phenotypes possessed bilateral, testicular gonads with associated epididymides and vasa deferentia. Female phenotypes possessed bilateral ovarian gonads with associated oviducts and uteri. Overt hermaphrodites possessed either an ovarian gonad and contralateral testicular gonad with respective accessory structures, or an ovarian or testicular gonad with a contralateral ovotestis associated with derivatives of both Mullerian and Wolffian ducts.

Histological Methods

Gonads were fixed overnight at room temperature in Bouin's fixative and stored in 70% ethanol. Conventional paraffin embedding and sectioning were performed and serial sections (6-8 um) were stained with hematoxylin and eosin.

Assay for Steroid Sulphatase

The assay for steroid sulphatase was a modification of previously published protocols[17-20]. Day 13-15 fetuses were sonicated with 15-20 pulses from a microtip attached to a Sonicator Cell Disruptor (Heat Systems-Ultrasonics, NY) in 0.5-1.0 ml 0.1 M ice-cold Tris buffer, pH 7.2. The sonicated tissues were microfuged for 10 min at 4° C. To 100 1 of the supernatant, 25 1 of thrice-extracted ^{3}H-dehydroepiandrosterone sulfate, ammonium salt (70,000-100,000 CPM, New England Nuclear, Mass.) was added and the reaction mixture was incubated for 4 hrs at 37°C in a shaking waterbath. The reaction was terminated with 100 1 of 0.1 N NaOH. 500 1 petroleum ether was added, the test tubes were vortexed then centrifuged (2000 rpm, 10 min). The aqueous phase was frozen on dry ice and the organic phase poured into scintillation vials and counted in 4 ml scintillation fluid (3a20, Research Products Internat. Corp., Ill.).

RESULTS

Backcrossing of B6.Y^{Dom} Overt Hermaphrodites to B6

Of the six overt hermaphrodites, five were fertile and capable of impregnating females. Despite all sires being overt hermaphrodites, 22% of the post-natal progeny possessed bilateral testicular gonads and were, thus, more masculinized than their fathers (see Table 1). Histological sections of the testicular gonads revealed developing oocytes either singly or in groups, in the rete testes or in neighboring seminiferous tubules, demonstrating that, despite their external appearance, the testicular gonads were ovotestes (Figure 1).

An examination of the ratios of post-natal phenotypes revealed that the B6.Y^{Dom} overt hermaphrodites generated phenotype ratios

Table 1: Progeny of five B6.Y^{Dom} overt hermaphrodites (#1-5) categorized into male, hermaphroditic and female phenotypes.

	Male (%)	Hermaphroditic (%)	Female (%)
#1	8 (25%)	4 (12%)	20 (69%)
#2	7 (24%)	2 (7%)	20 (69%)
#3	0	0	2 (100)
#4	8 (32%)	3 (12%)	14 (56%)
#5	1 (5%)	6 (27%)	15 (68%)
TOTAL	24 (21.8%)	15 (13.6%)	71 (64.5%)

not significantly different from those in our B6.Y^{Dom} colony for the N6 generation (male phenotypes = 47 (26%), hermaphrodite = 22 (12%), female phenotypes = 111 (62%); X^2 = 0.704). A significant increase in the incidence of female phenotypes was absent, suggesting that if other genes in the B6 genome play a role in generation of XY females, simply having these genes homozygous is not sufficient. We suspected that a gene or genes in the X-Y pairing/recombination region, which may be gained or lost during meiosis in the sires, may be involved in the regulation of phenotypes in *tda-1* sex reversal.

Matings of B6.Y^{Dom} to C3H/An and Backcrossing of (C3H/An x B6.Y^{Dom}) F1 Males to C3H/An and B6

The C3H/An mouse strain is deficient in steroid sulphatase activity[21,22]. By crossing C3H/An male mice to XO females, it was demonstrated indirectly that the murine *STS* locus exists on the X and Y chromosomes in the X-Y pairing/recombination region and that the locus undergoes an obligatory recombination during meiosis[20].

It can be postulated that a gene (or genes) in the X-Y pairing/recombination region is involved in *tda-1* sex reversal, if it can be shown that steroid sulphatase activity correlates with male, female or hermaphroditic phenotypes. Using this rationale, we crossed an N7 B6.Y^{Dom} to a C3H/An female, then backcrossed the (C3H/An x B6.Y^{Dom}) F1 males to both C3H/An and B6 females. We examined the fetal progeny for gonadal phenotype and steroid sulphatase activity. B6 and B6.Y^{Dom} strains possess steroid sulphatase activity. (C3H/An x B6) F1 fetuses (*STS* +/-) possess steroid sulphatase activity at a level which is intermediate between that of B6 (*STS* +/+) and C3H/An (*STS* -/-) fetuses (see Table 2). When (C3H/An x B6.Y^{Dom}) F1 are backcrossed to C3H/An females, progeny should be equally divided between *STS* +/- and -/- phenotypes, if there are no selective fetal deaths (see Figure 2). Similarly, when backcrossed to B6, progeny should be equally divided between *STS* +/+ and +/- phenotypes (see Figure 3). Assuming hermaphroditism is maintained in the various crosses, one can correlate male, female or hermaphroditic phenotypes with steroid sulphatase activity.

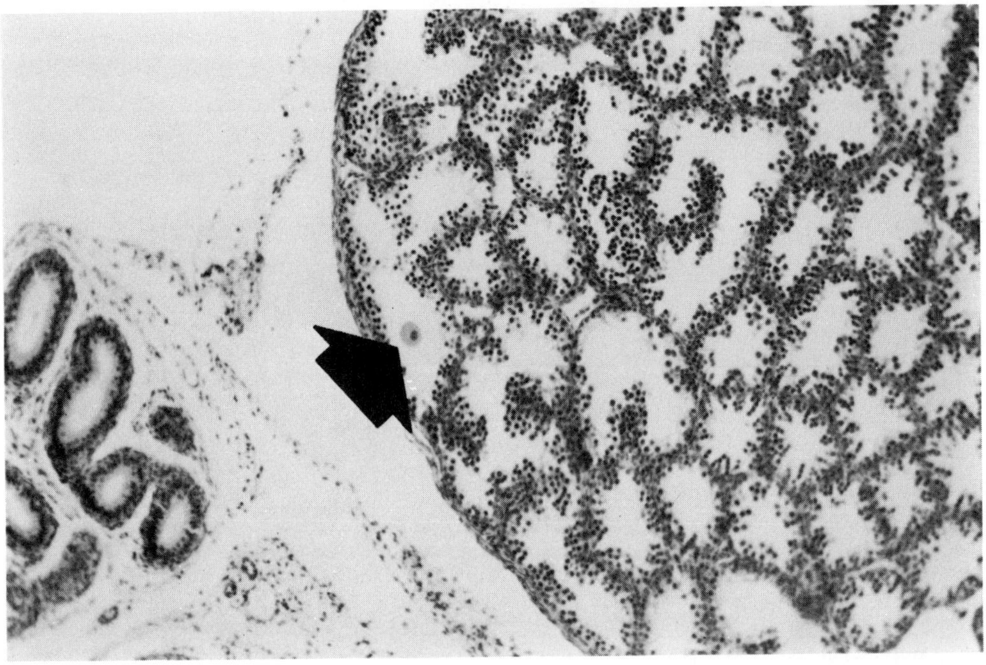

Figure 1: Histological section of a testicular gonad from an 18-day-old male phenotype whose sire was a B6.Y^{Dom} N5 overt hermaphrodite. A naked oocyte is present in a seminiferous tubule (arrow).

When B6.Y^{Dom} male phenotypes were crossed to C3H/An females, and the F1 fetuses examined on day 14/15 of gestation, the progeny consisted of males, females, and fetuses with aberrant testes similar to the ovotestes found in B6.Y^{Dom} [14]. Histological sections have recently revealed the presence of meiotic germ cells in the aberrant gonads, demonstrating that they were indeed ovotestes (data not shown). In mice, oocytes initiate meiosis at about day 15 of gestation while spermatocytes initiate meiosis only postnatally.

When (C3H/An x B6.Y^{Dom}) F1 males were backcrossed to C3H/An females, fetal progeny again consisted of male, female, and hermaphroditic phenotypes. Of 106 fetuses on day 14/15, 31 (29%) were males, 47 (44%) females and 28 (27%) were hermaphrodites[14]. We conclude that <u>tda-1</u> or <u>tda-1</u> related hermaphroditism can be transferred to C3H/An at least to the N2 generation. The absence of a sex ratio skewed in favor of females (X^2 = 1.36) suggests that XY females were either not present or, if present, were rare. For convenience, all fetuses possessing bilateral ovaries were assumed to be XX, and all fetuses possessing testes or ovotestes were assumed to be XY. When day 14-15 fetuses of all gonadal phenotypes were tested for steroid sulphatase activity, female and male phenotypes possessed similar numbers of <u>STS</u> -/- and <u>STS</u> +/- phenotypes. Hermaphrodities, however, showed a paucity of <u>STS</u> -/-

Table 2: Steroid sulphatase activity in fetuses of B6 (STS +/+), (C3H/An x B6) F1 (STS +/-), and C3H/An (STS -/-) in two separate experiments. Values are in CPM per 100 µl of supernatant of fetal homogenate (see Methods).

	Strain	n	Range	Mean + SD
Exp. 1	B6	4	66,030 - 90,520	77,340 ± 12,180
	(C3H/An x B6) F1	6	30,960 - 35,820	33,060 ± 1,910
	C3H/An	6	15,910 - 18,830	16,620 ± 1,110
Exp. 2	B6	4	29,100 - 30,940	29,910 ± 820
	(C3H/An x B6) F1	8	15,420 - 22,970	18,580 ± 2,660
	C3H/An	4	6,710 - 8,710	7,830 ± 830

phenotypes (see Figure 2, Exp. 1). When an additional 41 male and hermaphroditic fetuses were tested for steroid sulphatase activity, a similar trend was noted (see Figure 2, Exp. 2). Pooling the data from male and hermaphrodite steroid sulphatase activity from both experiments revealed that the group of hermaphrodites included significantly fewer STS -/- phenotypes (X^2 = 4.83, p = 0.05).

When (C3H/An x B6.Y^{Dom}) F1 males were backcrossed to B6, day 14 fetuses were either hermaphrodites or females. Normal males were not found. A statistically significant (X^2 = 6.8, p = 0.01) skewed sex ratio in favor of females suggests that either XY females were present or hermaphrodites were being lost. An analysis of steroid sulphatase activity of the fetuses revealed a significantly higher percentage of STS +/- phenotypes (X^2 = 4.24, p = 0.05) (see Figure 3). An examination of the data from the gonads revealed that while hermaphrodites were equally divided between STS +/- and +/+ phenotypes, 72% of the females (presumably XX and XY) were STS +/-, a value which was significantly different from the expected ratio of 1:1 (X^2 = 7.1, p = 0.01).

DISCUSSION

The data on the progeny of the overt hermaphrodite post-natal progeny support our previous suggestion that a genetic mechanism maintains the male, hermaphroditic and female phenotypes, as determined post-natally in each generation of our B6.Y^{Dom} colony[14]. The present data, as well as earlier data[14], do not support the hypothesis that homozygosity of a recessive B6 feminization gene is the major cause of XY female phenotypes. If such a gene were the major cause of XY females, one would expect the percentage of female phenotypes to increase with each backcross generation, since each backcross would result in more progeny being homozygous for the B6 feminization gene.

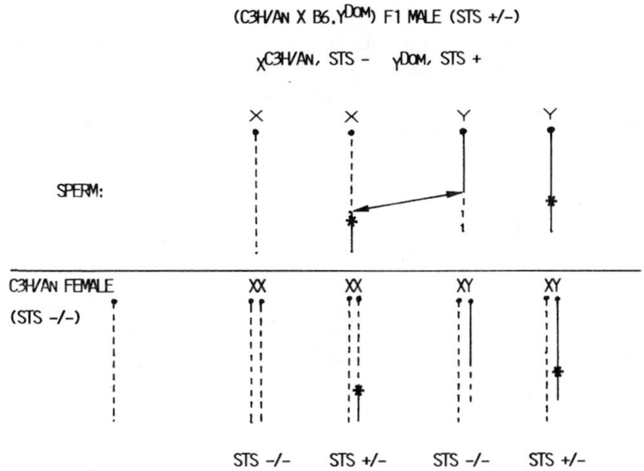

	STS -/-	STS +/-		STS -/-	STS +/-
EXPERIMENT 1	14	19	MALE	14	14
			HERM	3	10
EXPERIMENT 2	-	-	MALE	10	10
			HERM	8	13
		TOTAL:	MALE	24	24
			HERM	11**	23

** = HERMAPHRODITE STS -/- PHENOTYPE SIGNIFICANTLY DIFFERENT FROM MALE STS -/- PHENOTYPE (X^2 = 4.83, P = 0.05)

Figure 2: Steroid sulphatase activity in relation to gonadal phenotype in day 13-15 fetal progeny of C3H/An (STS -/-) x (C3H/An x B6.Y^{Dom}) F1 males (STS +/-). The obligatory pairing/recombination between one chromatid of the X, derived from C3H/An, and one chromatid of the Y, derived from *M. m. domesticus*, is marked by the double arrow. The locus of the steroid sulphatase gene is marked by an asterisk. The pairing/recombination between the STS deficient X^{C3H} and the STS normal Y^{Dom} results in four types of sperm. When backcrossed to STS-deficient C3H/An females, four types of progeny are formed. In such a cross, all fetuses with bilateral ovaries were assumed to be XX; all with ovotestes or testes were assumed to be XY. In Exp. 1, values for STS +/- fetuses ranged from 27,000-51,500 CPM; for STS -/- fetuses, 10,300-18,500 CPM. In Exp. 2, values for STS +/- fetuses ranged from 11,300-16,900 CPM; for STS -/- fetuses, 6,000 - 9,000 CPM. B6, (C3H/An x B6) and C3H/An fetuses were prepared and tested simultaneously as controls for STS +/+, +/-, and -/-, respectively, in both experiments. The three types of control gave non-overlapping values. The data demonstrate that the Y chromosome which did not undergo recombination is capable of giving rise to both males and hermaphrodites in this cross. The data also demonstrate that the Y chromosome which had undergone recombination gave rise to significantly fewer hermaphrodites than males.

We had suspected that Y-bearing sperm from B6.Y^{Dom} sires may possess slightly different Y chromosomes. One can envision that some Y-bearing sperm give rise to male phenotypes, others to hermaphrodites or XY females. A heterogeneous population of Y-bearing sperm can be formed if a gene (or genes) is gained or lost during X-Y pairing/recombination. A similar situation occurs in <u>Sxr</u>[9,23-25].

Our data from experiments using the <u>STS</u> locus as a marker for the X-Y pairing/recombination region, are suggestive that, indeed, the population of Y-bearing sperm may be heterogeneous, and that a gene (or genes) in the X-Y pairing/recombination region may be involved in regulating phenotype in <u>tda-1</u> sex reversal. In the C3H/An x (C3H/An x B6.Y^{Dom}) F1 cross, the <u>fetal</u> data reveal three major points. The first is that sperm which contain the Y which had not undergone recombination (<u>STS</u> +/-) gave rise to both male and hermaphroditic progeny (see Figure 2). How this occurs is not known. The second point is that equal numbers of <u>STS</u> -/- and +/- male phenotypes were formed. Thus, with respect to male phenotypes, there is no evidence for a selective advantage of the Y

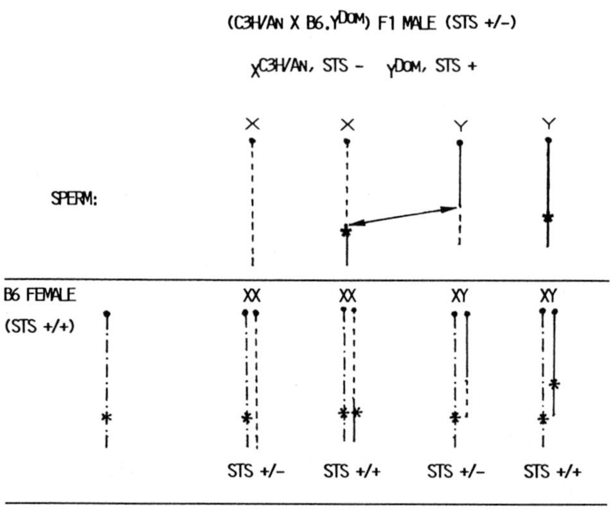

	STS +/-	STS +/+
FEMALES (XX + XY)	26**	10
HERMAPHRODITES (XY)	8	9

** = STATISTICALLY DIFFERENT FROM FEMALES STS +/+ (X^2 = 7.1, P = 0.01)

Figure 3: Steroid sulphatase activity in relation to gonadal phenotype in day 14 fetal progeny of B6 (<u>STS</u> +/+) x (C3H/An x B6.Y^{Dom}) F1 males (<u>STS</u> +/-). For B6 x (C3H/An x B6.Y^{Dom}) F1 fetuses, <u>STS</u> +/- was set at <18,600 CPM + 2 standard deviations (23,900 CPM). No F1 fetus was found to be <u>STS</u> -/-. All fetuses were either females (presumably XX or XY), possessing bilateral ovaries, or hermaphrodites (XY), possessing ovo-testes. The figure illustrates the progeny to be expected in relation to karyotype. Note that only the Y which had undergone recombination could give rise to <u>STS</u> +/- XY progeny.

which had not undergone recombination over the Y which had undergone recombination, with respect to fertilizing eggs and/or survival of the embryos. The third and most important point is that sperm that carry the Y chromosome which had undergone recombination gave rise to more than twice the number of male phenotypes than hermaphrodites. It is possible that a gene (or genes) in the X-Y pairing region centromeric to *STS* and associated with generation of hermaphrodites, when lost as a result of crossing over with the X, results in either non-viable sperm or embryonic/fetal death? Another possibility is that a gene from the X^{C3H}, when transferred to the Y chromosome, results in non-viable sperm or embryonic/fetal death in hermaphrodites but not in male phenotypes.

The B6 x (C3H/An x B6.Y^{Dom}) F1 cross also provided data that suggest that the X-Y pairing/recombination region may be involved in *tda-1* sex reveral and that the sperm, derived from the Y which had undergone recombination, may be abnormal. If the X-Y pairing/recombination region and the region which encodes steroid sulphatase activity were not involved in *tda-1* sex reversal, one would expect equal numbers of *STS* +/- and +/+ in hermaphrodites and in females, no matter what their karyotype. Indeed, equal numbers of hermaphrodites of both *STS* phenotypes were found. However, a large percentage of *STS* +/- females was observed. It is possibile that embryos or fetuses in the other categories died *in utero*.

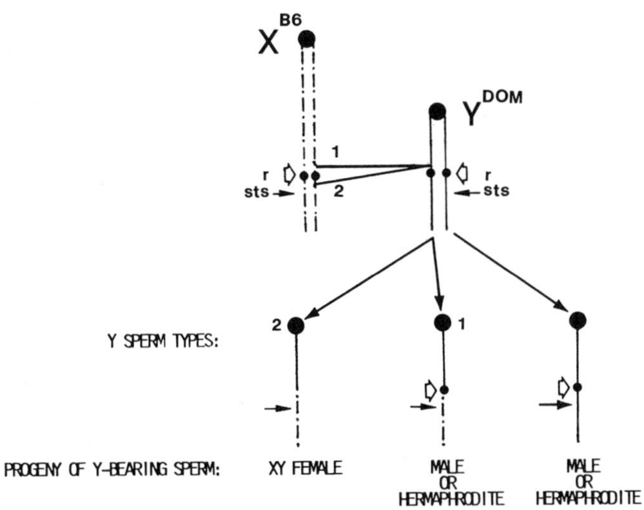

Figure 4: Diagram of the hypothetical mechanism for formation of XY females in the B6.Y^{Dom} strain. During meiosis in the male, an X and a Y chromatid undergo pairing/ recombination. A regulatory gene (r) in the X-Y pairing/ recombination region, exists centromeric of the *STS* locus (sts). If the Y chromatid maintains the regulatory gene, either as the result of nonrecombination or of recombination distal to the regulatory gene (1), the XY fetus differentiates into the male phenotype or overt hermaphrodite. If the regulatory gene is lost as in (2), the fetus differentiates into an XY female.

Our present hypothesis is that a regulatory gene (or genes) in the X-Y pairing/recombination region, centromeric to the STS locus, is involved in tda-1 sex reversal in the B6.Y^{Dom} strain (see Figure 4). When present, the Y^{Dom} chromosome is capable of inducing males and hermaphrodites. When this gene is lost from Y^{Dom} due to X-Y pairing/recombination, XY females may be formed or embryonic or fetal death occurs. Whether this gene is the tda-1 gene is presently not known. Experiments are in progress to provide further insight into the putative regulatory gene in the X-Y pairing region by use of recombinant DNA probes specific for the murine Y chromosome[26,27] and use of probes believed to be specific for the X-Y pairing/recombination region.

ACKNOWLEDGEMENTS

We thank Dr. Michael Potter for providing the stock M. domesticus under NCI contract N01-CB-25584; Ms. Joan Gedney, Elaine Sinclair-Thompson and Dr. Teruko Taketo for histological assistance; and Ms. Lisa Spatafora-Kalle for secretarial help. We especially wish to thank Dr. Ulrich Hammerling in whose laboratory this work was performed. This research was supported in part by grants from the National Institute of Child Health and Human Development HD-18067 and National Institutes of Health CA-08748.

REFERENCES

1. Gordon, JW, Ruddle, RH: Mammalian gonadal determination and gametogenesis. Science, 211:1265-71, 1981.

2. Wilson, JD, George, FW, Griffin, JE: The hormonal control of sexual development. Science, 211:1278-84, 1981.

3. Zamboni, L, Upadhyay, S: Germ cell differentiation in mouse adrenal glands. J Exp Zool, 228:173-93, 1983.

4. Upadhyay, S, Zamboni, L: Ectopic germ cells: Natural model for the study of germ cell sexual differentiation. Proc Natl Acad Sci USA, 79:6584-88, 1982.

5. McLaren, A: Studies on mouse germ cells inside and outside the gonad. J Exp Zool, 228:167-71, 1983.

6. Bonhomme, F, Catalan, J, Britton-Davidian, J, Chapman, VM, Moriwaki, K, Nevo, E, Thaler, L: Biochemical diversity and evolution in the genus Mus. Biochemical Genetics, 22:275-303, 1984.

7. Ferris, SD, Sage, RD, Prager, EM, Ritte, U, Wilson, AC: Mitochondrial DNA evolution in mice. Genetics, 105:681-721, 1983.

8. Gropp, A, Winking, H, Redi, C, Capanna, E, Britton-Davidian, J, Noack, G: Robertsonian karyotype variation in wild house mice from Rhaeto-Lombardia. Cytogenet Cell Genet, 34:67-77, 1982.

9. Eicher, E: Primary sex determining genes in mice. In: Amann, RP, Siedel, GE Jr, eds. Prospects for Sexing Mammalian Sperm. Colorado: Colorado Assoc. Univ. Press, 121-35, 1982.

10. Eicher, E, Washburn, LL, Whitney, JB III, Morrow, KE: Mus poschiavinus Y chromosome in the C57BL.6J murine genome causes sex reversal. Science, 217:535-37, 1982.

11. Eicher, E, Washburn, LL: Inherited sex reversal in mice: Identification of a new primary sex-determining gene. J Exp Zool, 228:297-304, 1983.

12. Gropp, A: Sex reversal in XY mice by a foreign Y-chromosome. In: Amann, RP, Siedel, GE Jr, eds. Prospects for Sexing Mammalian Sperm. Colorado: Colorado Assoc. Univ. Press, 136-138, 1982.

13. Nagamine, CM, Taketo, T, Koo, GC: Morphological development of the gonad in Tda-1 XY sex reversal. Differentiation (in press).

14. Nagemine, Cm, Taketo, T, Koo, GC: Studies on the genetics of tda-1 inherited sex reversal. (Submitted for publication).

15. Festing, MFW: Inbred Strains in Biomedical Research. NY: Oxford Univ Press, 1979.

16. Whitten, WK, Beamer, WG, Byskov, AG: The morphology of fetal gonads of spontaneous mouse hermaphrodites. J Embryol Exp Morph, 52:63-78, 1979.

17. Burstein, S, Dorfman, RI: Determination of mammalian steroid sulfatase with 7 a-H3-3 b-hydroxyandrost-5-en-17-one sulfate. J Biol Chem, 238:1656-60, 1963.

18. Shapiro, LJ, Cousins, L, Fluharty, AL, Stevens, RL, Kihara, H: Steroid sulfatase deficiency. Pediat Res, 11:894-97, 1977.

19. McNaught, RW, France, JT: Studies of the biochemical basis of steroid sulphatase deficiency: Preliminary evidence suggesting a defect in membrane-enzyme structure. J Steroid Biochem, 13: 363-73, 1980.

20. Keitges, E, Rivest, M, Siniscalco, M, Gartler, SM: X-linkage of steroid sulphatase in the mouse is evidence for a functional Y-linked allele. Nature, 315:226-27, 1985.

21. Balazs, I, Filippi, G, Rinaldi, A, Grzeschik, K-H, Siniscalco, M: Studies on human X-linked ichthyosis and steroid sulfatase in man, mice and their hybrids. Cytogenet Cell Genet, 25:133, 1979.

22. Balazs, I, Purrello, M, Rocchi, M, Rinaldi, A, Siniscalco, M: Is the gene for steroid sulfatase X-linked in man? An appraisal of data from humans, mice, and their hybrids. Cytogenet Cell Genet, 32:251-52, 1982.

23. Burgoyne, PS: Genetic homology and crossing over in the X and Y chromosome of mammals. Hum Genet, 61:85-90, 1982.

24. Hansmann, I: Sex reversal in the mouse. Cell, 30:331-32, 1982.

25. Singh, L, Jones, KW: Sex reversal in the mouse (Mus musculus) is caused by a recurrent nonreciprocal crossover involving the X and an aberrant Y chromosome. Cell, 28:205-16, 1982.

26. Bishop, CE, Boursot, P, Baron, B, Bonhomme, F, Hatat, D: Most classical Mus musculus domesticus laboratory mouse strains carry a Mus musculus musculus Y chromosome. Nature, (London) 315:70-2, 1985.

27. Lamar, EE, Palmer, E: Y encoded, species-specific DNA in mice: Evidence that the Y chromosome exists in two polymorphic forms in inbred strains. Cell, 37:171-77, 1984.

CENTROMERIC ALPHOID DNA ON THE Y CHROMOSOME

Phillip Kearney and Jonathan Wolfe

Department of Genetics and Biometry
University College London
Wolfson House
4 Stephenson Way
London NW1 2HE

The alpha satellite DNAs are a family of nucleotide sequences which are present in the genomes of all primate species so far examined[1,2]. They consist of tandem reiterations of a subunit of approximately 170bp. Variation in the subunit sequence can be observed both between different subunits isolated from a single species and between subunits isolated from different species. In humans, variation has also been observed in the way that the satellite DNA has been amplified at different sites in the genome. The majority of alphoid DNAs in lower primates consists solely of repeats of monomers or dimers of the basic subunit (as revealed by the presence of 'ladders' of restriction fragments, based on a step length of either 170bp or 340bp in Southern blots of genomic DNA). However, higher primates, including man, contain, in addition, numerous families of alphoid DNAs each of which gives a characteristic Southern restriction pattern and each of which shows different degrees of hybridization to cloned probes[3-10]. Alphoid DNA is predominantly located at the centromeric region of all chromosomes. Whether it serves a function there is unknown, but its consistent location, its conserved monomer length and the identification of a protein which binds specifically to the repeat[11] argue that it may indeed perform some role in the cell.

One of the most prominent families of alphoid DNA in the human genome can be seen after digestion of genomic DNA with EcoR1 and electrophoresis in agarose, as a 340bp, ethidium bromide-stained, fragment. This simple family of dimers has been studied extensively by Manuelidis and her colleagues[12,13] and has been shown to resemble the lower primate, alphoid satellites. However, in the course of primate evolution many, more elaborate arrangements of alphoid DNA have occurred. Digestion of human DNA with various restriction endonucleases and Southern hybridization to cloned probes has revealed that, superimposed on the 170bp subunit repeat structure, there are higher order periodicities which reflect the saltatory amplification of blocks of contiguous subunits. Recent findings are consistent with the hypotheses that each human chromosome carries its own family of alphoid DNAs, which has diverged in sequence from the average sequence of the whole genome and which has its own, higher order periodicity. In this paper, we compare published data about the sequences elsewhere in the genome with data derived from the alphoid DNA on the Y chromosome. We also show for the first time that the amount of alphoid DNA on the human Y chromosome is highly variable.

Variations in Sequence

A large number of alphoid sequences has been cloned. Some sequences have been assigned to a specific chromosome, either by consideration of the source from which the DNA was cloned or from genetic evidence, and the DNA from some of the clones has been sequenced. Of the clones whose DNA has been sequenced, every clone is different and none of the sequences corresponds exactly to the consensus sequence of the uncloned 340bp EcoRl fragment, published by Wu and Manuelidis[13]. No systematic approach has been adopted to the problem of obtaining clones from every chromosome. However, the piecemeal method has so far produced chromosome-specific clones from roughly half of the human chromosomes (refs 3-5, 8-10, 14, and unpublished results from several laboratories).

Jorgensen, Bostock and Bak[10] have approached the problem in two ways. In a study of 45 clones, which contained alphoid DNA isolated from the EcoRl 340bp and 680bp fragments of human DNA, they found that a number of subfamilies could be distinguished (on the basis of sequence) within which members were more closely related to each other than to the consensus sequence. Their second approach was to isolate clones from a somatic cell hybrid in which chromosome 7 was the only human chromosome. Southern blotting of DNA from this hybrid revealed that the organization of alphoid DNA on chromosome 7 is probably based on a repeat unit of eight dimers of 340bp. Digestion with BamHl gives only a single band of 2.7Kb (consistent with sixteen subunits), whereas digestion with EcoRl gives principal bands that correspond to dimers, tetramers and hexamers (suggesting a structure based on dimers). They sequenced DNA from seventeen clones derived from the 340bp and 680bp bands produced by digestion with EcoRl. All sequences were different, but together they form a closely related family whose consensus sequence have diverged by only 11% from the consensus sequence of Wu and Manuelidis but whose members have on average only diverged by 1.8% from each other.

Two groups have cloned representative members of the X chromosome-alphoid DNA and shown this satellite to have a higher order repeat structure characterized by the occurrence of a BamlH site every 2.0Kb[3,5]. DNA from one clone has been sequenced, revealing for the first time the relationship between the twelve subunits of 170bp of which it is comprised[15]. The twelve monomers are between 64% and 85% identical in sequence to each other but their consensus sequence is 80% and 84% homologous to the two halves of the consensus sequence of Wu and Manuelidis. Consideration of the internal homologies in the repeats on the X chromosome led Waye and Willard to propose that genesis of the 2.0Kb unit may have involved several steps. First, amplification of a monomer followed by sequence divergence, and subsequently amplification of a hexameric unit followed by further amplification and sequence divergence. However, they point out that because we do not know what role might have been played by other mechanisms of sequence rectification, this sequence of events is only one possible pathway among many.

We have obtained some sequence data from the alphoid DNA component of the Y chromosome[8]. The higher order repeat unit on this chromosome is revealed on Southern blots as a 5.5Kb EcoRl fragment. There is no evidence of any substructure of dimers, and it seems likely that a block of about 32 monomers has been amplified on this chromosome. Nine monomers from a clone which carries a single copy of the 5.5kb repeat unit have been sequenced. The degree of sequence divergence between the monomeric units on the X chromosome suggests that it is unwise to interpret these incomplete data definitively. Nevertheless, it appears that the Y chromosome consensus sequence may be more closely related to the consensus sequence derived from studies of the X chromosome (87% homologous) than to either of the two halves of the consensus sequence of Wu and Manuelidis (83% and 80% homologous). The divergence in nucleotide sequence is sufficient to explain why early experiments, which included hybridization _in situ_ of uncloned alphoid dimers, failed to demonstrate hybridization to the Y chromosome.

Human DNA digested with Xba I reveals a satellite of 682bp, which is another diverged alphoid sequence[6]. DNA from a clone of this sequence hybridized at high stringency only to human, chimpanzee and gorilla DNA and not to orangutan or gibbon DNA. Its chromosomal location has not been established. These data suggest that this satellite is probably derived from an alphoid DNA which has been amplified relatively recently.

Alphoid DNA is found as a higher proportion of total DNA in small, circular, extrachromosomal molecules than in the genome as a whole[16]. Two such molecules have been cloned and sequenced[16]. They exhibit considerable divergence both from the consensus sequence of Wu and Manuelidis and from each other. It is tempting to speculate that these small, circular molecules represent accidental by-products of the processes of amplification and homogenization of sequences. They could very easily be generated by recombination between repeat units on the same strand of DNA.

Table 1 aligns all the relevant published sequences, with the exception of the 45 random clones described in reference 10. Where possible, we have aligned consensus sequences. It is noteworthy that some regions of the sequence are more prone to divergence than others. Conversely, some regions, including the three alphoid protein binding sites[11], are conserved.

Organization of sequences on individual chromosomes

We have already discussed clones derived from chromosomes 7, X and Y. Several additional clones have been assigned to individual chromosomes. Table 2 lists the available information. The alphoid nature of the 3Kb Bam H1 fragment which is repeated on chromosome 6 has yet to be conclusively proved[4]. However, DNA from the cloned probe has been shown to hybridize to a 2kb BamH1 fragment, which is tandemly repeated on the X chromosome and which is presumably the alphoid repeat. Problems with cross hybridization of probes to alphoid DNA from several chromosomes have made it difficult to assign clones generated from sources which contain more than a single human chromosome. Non-reciprocal exchange occurring between chromosomes could also lead to the presence of identical satellites on more than one chromosome. For instance, the alphoid clone L1.26 constructed by Devilee et al[9] hybridizes strongly both to chromosomes 21 and 13 and it is not yet possible to prove from which chromosome it is derived.

Of the sequences displayed in Tables 1 and 2, only two, the Xbal satellite[6] and the chromosome 7 alphoid repeat[10], are organized in

TABLE 1

```
                                            I
                                        +-------+
                     10        20       :30     :  40        50
340bp   I    AATTCTCAGTAACTTCCTTGTGTTGTGTGTATTCAACTCACAGAGTTGAA
EcoR1   II   G        A    C T     A      CG                T

682bp   I    C        G      T     A   T  C     CG       AC
Xba1    II   C        A     AT   C A       CC      A
        III  C        A       C    A   T  C          T
        IV   C        A            A        C  G G A

chr.7   I                A   T            C                G
        II   C        A  A   T     A      CG          T    T

chr.X        C        A      T     A  AT  C

chr.Y        C     G  A      T     A      N     T

pc21         CT    G  A      T     A      CT    T

Sau3A1       C        A      T     A  AT  C


                     60        70        80        90       100
340bp   I    CGATCCTTTACACAGAGCAGACTTGAAACACTCTTTTTGTGGAATTTGCA
EcoR1   II    CT T    T  T       TTAG          G     AA G C

682bp   I     AT   C  T  T      TGT              C   A T  C A
Xba1    II    CT T    TG T C A  TTT GC               A    C
        III   AC T CC T GT C     GT                  AAC     G
        IV    CT      TG         TT     G            A    C

chr.7   I      TC      G          T                 C
        II    CT T    T  T       TT G          G     AA G C

chr.X         AT      TG T       TT N                A    C

chr.Y         CT T    TG TN      TT                  A    C N

pc21          CT T    TG TT      TT G      G         A    C

Sau3A1        AT    N NG T       NT  N    N          A    C
```

TABLE 1 (Continued)

```
                         II                                  III
                      +-------+                            +-------+
                      :  110  :    120       130       140  :    150
340bp       I     AGTGGAGATTTCAGCCGCTTTGAGGTCAATGGTAGAATAGGAAATATCTT
EcoR1       II          T   CAGA  T        C TTC T G   AC  G T -

682bp       I      AC   C      AA      C   C T     GAG A
Xba1        II          T    G ATA       A  GTT C  T G AC  G
            III         C    GCAG          C T   T GA  A
            IV          T    TGATAC        ATTTC T G  CAC  G-     G

chr.7       I                 TAG CA     T C   CA      AG
            II          T  A GGA  G A      C TTC T G   AC       -

chr.X                   T    GGA  N        NNTTC T G   AC  N

chr.Y                   T    GGAG CT       C T     G   A

pc21               AN   T    GTNGNC        C T     GA  A

Sau3A1                  N    GGA  N        C TTCNT N   AC  G   N N

                         160       170
340bp       I     CCTATAGAAACTAGACAGAAT
EcoR1       II     A   TATG-          G

682bp       I      AA   A    C        G
Xba1        II     A    A  TCA        G
            III    TCC  A    C        G
            IV     A    A  T      G   G

chr.7       I      AA   A    C
            II     A T CATG-          G

chr.X              AC   A       A     G

chr.Y              AC   A       C     C

pc21               AC   A             G

Sau3A              AC N A       N  G  G
```

Legend to Table 1

340bp: consensus sequence of ref. 13, two monomers;
682bp: sequence of ref. 6, four monomers;
chr. 7: consensus sequence of ref. 10, two monomers;
chr. X: consensus sequence of ref. 15;
chr. Y: consensus sequence of ref. 8;
pc21: consensus sequence of clone pc21, ref 16;
Sau3A1: consensus sequence of ref. 17.

The footprints of the alphoid binding protein are marked above the sequences, as I, II and III. (ref 11).

dimers, like the archetypal sequence of Wu and Manuelidis. It is quite possible that this dimeric organization is a more uncommon arrangement of alphoid DNA than has hitherto been supposed.

Restriction fragment length polymorphisms of alphoid DNA

It is in the nature of satellite DNA to be polymorphic. One has only to observe the range of variations in restriction sites and DNA sequences between the cloned alphoid sequences discussed above to realize that polymorphism is likely to be characteristic of the alphoid satellites too. The availability of cloned probes which hybridize specifically to only the alphoid DNA of a single chromosome has made it possible to demonstrate the presence of polymorphic restriction fragments, unobscured by irrelevant, cross-hybridizing bands derived from other chromosomes.

Willard et al[14] have used a subclone derived from the 2Kb BamH1 repeat on the X chromosome, which, under stringent conditions, gives an entirely X-specific pattern of hybridization. By digesting with restriction enzymes which do not cut the X chromosome consensus sequence but which do cut infrequent variants, they can visualize a 'ladder' of hybridizations to multimers of the 2Kb fragment. Individuals vary in terms of the presence and intensity of 'rungs' on the ladder and a very large number of haplotypes exists. The polymorphism information content[18] of this type of analysis is very high, principally because a very large number of restriction enzyme sites is being investigated simultaneously with each digest.

A second approach to analysis of polymorphisms is to digest genomic DNA with an enzyme which is known to cut once per higher order repeat. Jabs et al[4] showed that individuals were polymorphic for minor Bam H1 bands when hybridized to p308, a clone derived from the alphoid sequence specific for chromosome 6. The variant bands segregated in a Mendelian fashion but whether they were linked and whether they were also on chromosome 6 was not established. Willard et al[14] have used a similar approach

Table 2

Chromosome	Repeat Size	Diagnostic enzyme	Organization	Reference
6	3Kb	BamH1	ND	4
7	2.7Kb	BamH1	8 x 340bp	10
17	2.7Kb	EcoR1	16 x 170bp (plus some 15-mers and 14-mers)	14
18	ND	ND	ND	9
X	2.0Kb	BamH1	12 x 170bp	3,5,15
Y	5.5Kb	EcoR1	32 x 170bp	8

ND = not determined

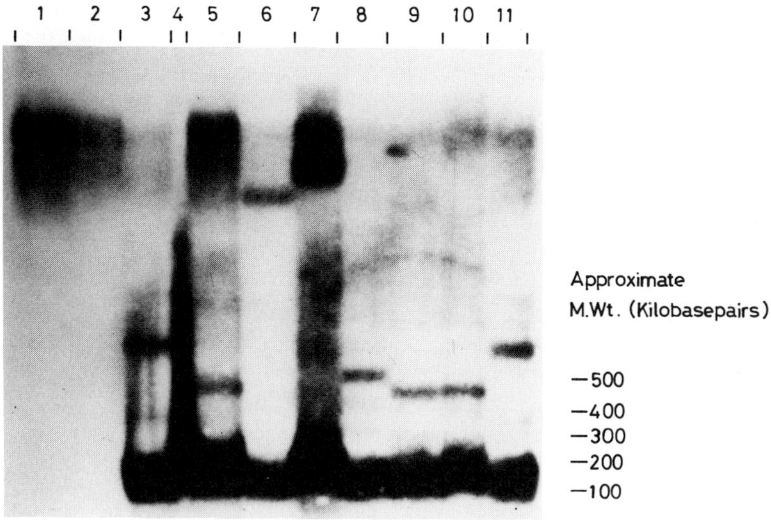

Figure 1: High molecular weight DNA was prepared from white blood cells, embedded in agarose blocks (2.5 x 10^5 cells per ml), essentially as described[24], by prolonged digestion in 0.5M EDTA pH9.0, 1% Na lauryl sarkosine, with 2 mg per ml Proteinase K at 50°C. The blocks were stored in 50% glycerol, 100 mM EDTA at -20°C. Prior to digestion, the blocks were washed extensively in 10 mM TrisHCl pH7.5, 1mM EDTA at 50°C. Digestions with BamH1 were carried out in a volume of 200 μl of medium salt buffer[25] with a twenty-fold excess of enzyme, for three hours. The blocks were transferred to the wells of a 1% agarose gel (Biorad DNA grade) in a Pharmacia 200 gel tank, in which the electrodes had been replaced by platinum wires. Electrophoresis was carried out at 250 volts for 15 hours in 0.5 x TBE buffer which was continuously recirculated through a cooling water bath. During the course of the electrophoresis, the polarity of the electrodes was continuously switched by a relay controlled by a BBC B+ microcomputer. Details of the circuitry are available from the authors. The computer was programmed to give a ratio of three steps forward steps to one step backward step, and the step length was continuously varied in a linear progression from 1 second to 20 seconds over the course of the run. After acid depurination, alkali denaturation and equilibration in 25mM sodium phosphate buffer pH 7.2, the DNA was electroblotted onto an Amersham Hybond nylon membrane. Hybridization of nick translated cY84 probe8 was carried out at 37°C in 50% formamide, 3 x SSC, 1 x Denhart's buffer, 50 μg per ml herring sperm DNA and the filter was washed to a final stringency of 0.2 x SSC at 65°C . The

(continued)

Figure 1 (Continued)
autoradiograph was exposed for 3 days at -70°C with an intensifying screen. Lanes 1 and 2: undigested controls; Lane 3: DNA from male 1, digested with Bam H1; lane 4: bacteriophage lambda DNA oligomers (overexposed), lanes 5-10: DNA from males 2 to 6, digested with Bam H1 lane 11: repeat of lane 3

with a probe which is highly specific to chromosome 17. In this case, alleles could be shown to be linked and no recombination was observed.

A third method is to digest genomic DNA with a restriction enzyme whose site of cleavage is not present in the consensus sequence from a single chromosome, and which cannot easily be created by mutation. Restriction fragments can then be separated by either orthogonal field agarose gel electrophoresis (OFAGE)[19] or by field inversion gel electrophoresis (FIGE)[20] and analyzed by Southern blotting.

We have digested samples of genomic DNA with Bam H1 and, after separation by FIGE, have probed blots with DNA from cY84, a clone which contains three repeat units derived from the Y centromeric alphoid satellite. Figure 1 shows that the population is highly polymorphic for the size of the single restriction fragment that hybridizes to cY84. In the first seven males analyzed, five different restriction fragments were found which ranged from 200 kb to greater than 1 Mb. If the entire length of the restriction fragment is assumed to be composed of tandem repeats of alphoid DNA (a reasonable assumption given the relative frequency of Bam HI sites elsewhere in the genome), this result corresponds to a variation in copy number of the repeat of from approximately 40 to 200 copies. The number of copies on the Y chromosome of the cell line OXEN is approximately 2008 and this value correlates with the band size, seen on FIGE, of approximately 1 Mb. (experiment not shown). Experiments to demonstrate correlation of quantitative variation in Y alphoid DNA to band size after FIGE are in progress.

We have also examined DNA from all members of a family which included a mother, a father and eight sons (results not shown). No variation in band size was observed between individual males.

Short stretches of tandemly repeated DNA, called mini-satellites, have been shown to be highly polymorphic as a result of variation in the number of copies of the repeat unit[21]. Models of satellite amplification and sequence homogenization, which invoke unequal recombination, predict that such copy number polymorphism should be found in any array of tandemly repeated DNA sequences. We find extensive polymorphism, presumably of copy number, on the Y chromosome but this polymorphism cannot be caused by unequal recombination between chromosomes because, with the exception of the distal region of Yp, which undergoes homologous recombination[22,23] with the tip of Xp, the chromosome is present in only one copy per diploid nucleus. Unequal recombination, if it occurs must be between sister chromatids.

The Y chromosome provides a particularly interesting source of material with which to study alphoid DNA. Despite (or perhaps because of) the relative isolation of the Y centromere from the rest of the genome, an alphoid repeat structure has evolved that is unique to this chromosome. Further experiments to investigate the copy number polymorphism will be aimed at elucidating how this process continues.

REFERENCES

1. Singer, MF: Highly repeated sequences in mammalian genomes. Int Rev Cytol, 76:67-111, 1982.

2. Donehower, L, Gillespie, D: Restriction site periodicities in highly repetitive DNA or primates. J Mol Biol, 134:805-834, 1979.

3. Yang, TP, Hansen, SK, Oishi, KK, Ryder, OA Hamkalo, BA: Characterization of a cloned repetitive DNA sequence concentrated on the human X chromosome. Proc Natl Acad Sci USA, 79:6593-6597, 1982.

4. Jabs, EW, Wolf, SF, Migeon, BR: Characterisation of a cloned DNA sequence that is present at the centromeres of all human autosomes and the X chromosome and shows polymorphic variation. Proc Natl Acad Sci USA, 81:4884-4888, 1984.

5. Willard, HF, Smith, KD, Sutherland, J: Isolation and characterisation of a major tandem repeat family from the human X chromosome. Nucleic Acids Res, 11:2017-2033, 1983.

6. Gray, KM, White, JW, Costanzi, C, Gillespie, D, Schroeder, WT, Calabretta, B, Saunders, GF: Recent amplification of an alpha satellite DNA in humans. Nucleic Acids Res, 13:521-535, 1985.

7. Willard, HF: Chromosome specific organisation of human alpha satellite DNA. Am J Hum Genet, 37:524-532, 1985.

8. Wolfe, J, Darling, SM, Erickson, RP, Craig, IW, Buckle, VJ, Rigby, PWJ, Willard, HF and Goodfellow, PN: Isolation and characterization of an alphoid centromeric repeat family from the human Y chromosome. J Mol Biol, 182:477-485, 1985.

9. Devilee, P, Cremer, T, Slagboom, P, Bakker, E, Scholl, HP, Hager, HD, Stevenson, AFG, Cornelisse, CJ, Pearson, PL: Two subsets of human alphoid repetitive DNA show distinct preferential localization in the pericentric regions of chromosomes 13, 18 and 21. Cytogenet Cell Genet, 41:192-301, 1986.

10. Jorgensen, AL, Bostock, CJ, Bak, AL: Chromosome-specific subfamilies within human alphoid repetitive DNA. J Mol Biol, 187:185-196, 1986.

11. Strauss, F, Varshavsky, A: A protein binds to a satellite DNA repeat at three specific sites that would be brought into mutual proximity by DNA folding in the nucleosome. Cell, 34:889-901, 1984.

12. Manuelidis, L: Chromosomal localisation of complex and simple repeated human DNAs. Chromosoma, (Berl.) 66:23-32, 1978.

13. Wu, JC, Manuelidis, L: Sequence definition and organisation of a human repeated DNA. J Mol Biol, 142:363-386, 1980.

14. Willard, HF, Waye, JS, Skolnick, MH, Schwartz, CE, Powers, VE and England, SB: Detection of restriction fragment length polymorphisms at the centromeres of human chromosomes by using chromosome-specfic alpha satellite DNA probes: Implications for development of

centromere-based genetic linkage maps. Proc Natl Acad Sci USA, 83:5611-5615, 1986.

15. Waye, JS, Willard, HF: Chromosome specific alpha satellite DNA: Nucleotide sequence analysis of the 2.0 kilobase pair repeat from the human X chromosome. Nucleic Acids Res, 13:2731-2743, 1985.

16. Jones, RS, Potter, SS: Characterisation of cloned human alphoid satellite with an unusual monomeric construction: evidence for enrichment in HeLa cells of small polydisperse circular DNA. Nucleic Acids Res, 13:1027-1042, 1985.

17. Kiyama, R, Matsui, H, Oishi, M: A repetitive DNA family (Sau 3A family) in human chromosomes: Extrachromosomal DNA and DNA polymorphism. Proc Natl Acad Sci USA, 83:4665-4669, 1986.

18. Botstein, D, White, RL, Skolnick, M, Davis, RW: Construction of a genetic linkage map in man using restriction fragment length polymorphisms. Am J Hum Genet, 32:314-331, 1980.

19. Schwartz, DC, Cantor, CR: Separation of yeast chromosome-sized DNAs by pulsed field gradient gel electrophoresis. Cell, 37:65-75, 1984.

20. Carle, GF, Frank, M, Olson, MV: Electrophoretic separations of large DNA molecules by periodic inversion of the electric field. Science, 232:65-68, 1986.

21. Jeffries, AJ, Wilson, V, Thein, SL: Hypervariable 'minisatellite' regions in human DNA. Nature, 314:67-73, 1985.

22. Cooke, HJ, Brown, WRA, Rappold, G: Hypervariable telomeric sequences from the human sex chromosomes are pseudoautosomal. Nature, 317:687-692, 1985.

23. Simmler, MC, Rouyer, F, Vergnaud, G, Nystrom-Lahti, M, Ngo, KY, de la Chapelle, A, Weissenbach, J: Pseudoautosomal DNA sequences in the pairing region of the human sex chromosomes. Nature, 317:692-697, 1985.

24. Gardiner, K, Laas, W, Patterson, D: Fractionation of large mammalian DNA restriction fragments using vertical pulsed-field gradient gel electrophoresis. Somatic Cell and Mol Genet, 12:185-195, 1986.

25. Maniatis, T, Fitsch, EF, Sambrook, J: Molecular Cloning: A Laboratory Manual, Cold Spring Harbor Laboratory Press, Cold Spring Harbor, 1982.

DO BKM SEQUENCES PLAY A ROLE IN HUMAN SEX DETERMINATION?

Robert P. Erickson, Arturo Bevilacqua, Craig Ross, Scott Donaldson, and John R.D. Stalvey

Departments of Pediatrics and Communicable Diseases and Human Genetics
The University of Michigan
Ann Arbor, Michigan 48109

In 1976, Singh et al. reported that female DNA of some species of snakes contains an extra satellite band that is absent from male DNA[1]. This satellite DNA from the Banded krait (Bkm, Banded krait minor) was later isolated and used as a probe in a number of studies[2]. When a *Drosophila* library was screened with Bkm, several strongly cross-hybridizing plasmids were detected, and these plasmids have been used interchangeably with the Bkm satellite DNA[3]. (We have used pCS316, one of these *Drosophila* clones, for many of our studies.) Both probes cross-hybridizes with DNA from a great variety of animals, including mammals and birds, and the hybridization studies suggest either marked sequence conservation[2,3] or independent evolution of the sequence[4]. The Bkm homologous sequences appear to be more abundant, or to be arranged in a distinct manner in the heterogametic sex in these species, just as it is in the Banded krait. Although Bkm probes demonstrate marked differences in the restriction patterns of male and female mouse DNA, such differences have not been found in male and female human DNA even though it has been estimated that 0.5-1% of total human DNA is related to the Bkm sequence[3].

Sequence analysis and expression studies of Bkm-related probes indicate that they may include structural genes[5,6]. DNA sequencing studies have revealed many repeats of the tetranucleotide GATA, in open reading frames which could code for the amino acid sequence leucine-serine-isoleucine-tyrosine (repeats of GACA are also found). Northern analyses of liver mRNA using Bkm probes have been reported to detect male-specific mRNAs[5,6] but Epplen's group did not find sex-specific differences in hybridizing mRNAs when only GATA and GACA repeats were used as probes[5,7]. However, it remains possible that a very hydrophobic protein encoded by multiple repeats of GATA is expressed, and that its expression could be involved in the development of the male gonad. If Bkm sequences are involved in sex determination, a stronger argument might be made that they serve as a control element, either unexpressed or expressed. For instance, at one time it seemed possible that Bkm sequences might play a regulatory role in the expression of many genes involved in sexual differentiation, analogous to the role proposed by Sutcliffe et al[8,9] for the ID sequences they reported in populations of mRNA from rat brain. However, the concept and existence of such identifier sequences has been challenged[10,11]. An example of such a shared potential control sequence is the "homeo box" found in several homeotic genes in *Drosophilia*. Homeotic mutations are thought to identify genes that specify the development of different parts of *Drosophila* larvae - their domains of action are known as developmental compartments. The conserved "homeo box" sequence, which can be present as several copies in each homeotic complex, is transcribed and is believed to encode a DNA-binding protein that may have some regulatory role in control of all these developmentally important genes[12,13]. As with the Bkm sequences, the "homeo box" sequences are highly conserved, and are found in vertebrates[14] where they are also expressed[15,16,17]. The Bkm sequences, if not active in a manner analogous to that of homeo boxes, could function by binding regulatory proteins, as suggested by Chandra[18].

It remains to be determined whether Bkm sequences are involved in sex determination. This chapter presents a review of our studies of Bkm in *Homo sapiens*.

Quantitation of Bkm sequences in man, compared to other vertebrates

Although it has been reported that Bkm-related sequences "amount to 0.5-1% of the total DNA"[3], the data and basis for this estimate were not given. In an initial screen of about 100 human Y chromosomal cosmids we were not able to detect any that cross-hybridized to the *Drosophila* Bkm probe, pCS316[19]. We wondered whether Bkm sequences in the salmon sperm DNA, which was used as carrier during the hybridization, were competing with any hybridization to the cosmids. Thus, we performed quantitative studies of the hybridization of Bkm sequences to the DNA of a variety of vertebrates, using *E. coli* DNA as carrier. Standard optical density methods are not satisfactory for the quantitation of DNA in the presence of even small amounts of protein or RNA, so the DNA in our experiments was quantitated by a sensitive, DNA-specific chemical procedure[20]. (The popular method of quantitating DNA by measurement of hybridization to a highly repetitive repeat, such as *Alu* for human DNA, is not possible in this cross-species situation.) As seen in panel A of Figure 1, *E. coli* DNA did not hybridize to the probe (when *E. coli* DNA was used as carrier) while salmon sperm DNA showed significant hybridization. *Drosophila* DNA (Oregon R) gave strong signals, as did DNA from mice, but DNA from the higher primates showed much less hybridization. Presumably the cross-species hybridizations are due to the shared GATA/GACA sequences. In the case of *Drosophila*, the GATA/GACA as well as the non-Bkm sequences in the probe, could have contributed to the signal. Thus, the same DNAs were also hybridized to the synthetic oligodeoxynucleotide $(GATA)_5$. As seen in panel B of Figure 1, relatively similar patterns of hybridization were obtained, even when *Drosophila* DNA was used.

We quantitated the extent of hybridization on both sets of filters using a Zeineh Soft Laser scanning densitometer. The readings were only linear to about 200 area units, so values greater than this, when found at the lowest amount of DNA (0.5 g), represent the lower limit of accuracy. These semi-quantitative results are shown in Table I. *Drosophila* DNA hybridized more strongly than salmon DNA, while mouse DNA gave even stronger signals with either probe. We studied a variety of mice with varying amounts of Y material on the 129/SvPas background. Sxr is a fragment of the Y chromosome which is male-determining and contains Bkm sequences in the male specific organization detected by certain restriction enzymes[21]. Thus, the various genotypes of 129/SvPas mice have been arranged from top to bottom of Table 1 in terms of increasing amounts of Y material: no Y chromosome (2 individual XX females), a Y fragment (XX, Sxr male) 1 Y chromosome (an XY male), and a Y chromosome plus the Sxr fragment (a XY, Sxr carrier male). No clear trend of increased hybridization with either probe is seen. Furthermore, DNA from an A/HeJ sample of unknown sex (but certainly not "2" Y) gave as high values as DNA from the XY, Sxr carrier. Thus, we cannot quantitatively distinguish male and female mouse DNA with either probe; a result also obtained by others[2].

Regardless of which probe is used, the DNA of higher primates does not hybridize to Bkm-related sequences to nearly as great an extent as does that of mice. The fact that DNA from higher primates gives similar degrees of hybridization with the 2 probes (under these conditions), while DNA from mice and *Drosophila* gives roughly two-fold greater signals

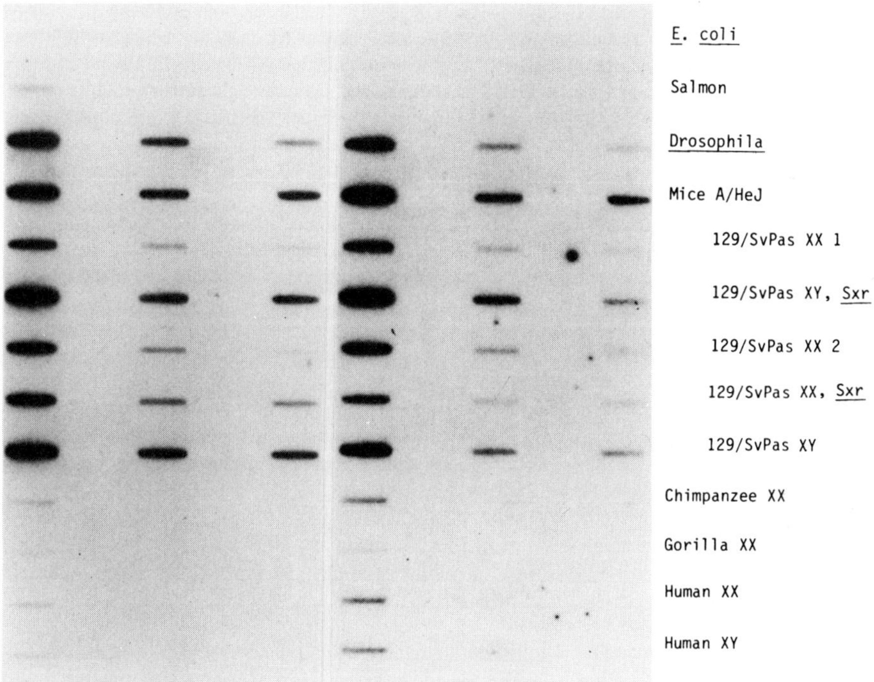

Figure 1: Hybridization of Bkm-related pCS316 (panel A) or $(GATA)_5$ (panel B) to DNA from a variety of vertebrates (and the *E. coli* DNA control). Methods: DNA was quantitated by the fluorometric diaminobenzoic acid assay[20]. In each panel, 5 μg, 1 μg, and 0.5 μg of the indicated DNAs have been applied left to right. The nick-translated Bkm-related insert from pCS316 (10^8 cpm/μg) was hybridized in 1 M NaCl, 1% SDS, 50% formamide, and 25 μg/ml denatured *E. coli* DNA at 42°. The filter was washed with 0.5x SSC at 65°C. The $(GATA)_5$ was synthesized on a Beckman oligonucleotide synthesizer and end-labelled with T4 polynucleotide kinase. 10^7 cpm were hybridized to a similar filter in 1 M NaCl, 1% SDS, with 30 μg/ml denatured *E. coli* DNA at 10^6 cpm/ml 45°C. The filter was washed twice in 6x SSC at 45°C for 30 min.

with pCS316 than with $(GATA)_5$, may reflect the presence of GACA repeats together with GATA repeats in pCS316. Thus, salmon DNA, with a 7-fold greater level of hybridization to pCS316 than to $(GATA)_5$, ought to be particularly rich in GACA repeats.

We have repeated our screen of the original 100, Y chromosomal cosmids[19], and screened 30 more cosmids generated by the same methods (Jonathan Wolfe, unpublished), with the pCS316 insert in the presence of *E. coli* DNA as carrier, and we have still not detected positive cosmids. However, others have found Bkm-positive, Y chromosomal cosmid clones[23] as would be predicted from our data from chromosomal hybridization *in situ* (see below).

Chromosomal locations of Bkm-related sequences in man

In an initial study, we determined the location of Bkm sequences on mouse autosomes as well as on the X and the Y chromosomes[24]. Using pCS316, we found two regions of increased concentration on the X chromosome and regional concentrations on two autosomes, one of them being proximal 17 in a region where deletions can cause hermaphroditism[25]. Our studies on the localization of Bkm sequences on human chromosomes provided contrasting results. Although humans have decreased amounts of Bkm-hybridizing sequences, sequences detected with pCS316 are widely scattered, with some evidence of hybridization on all chromosomes[26]. Regions of concentration were found on the human X at p21 and q21 as well as on autosomes six and eleven[26]. However, in contrast to what has been found in mice, there is no strong regional localization of Bkm sequences on the human Y chromosome[26].

The relative lack of Bkm-related sequences on the human Y chromosome is striking, since localization of Bkm sequences on the Y chromosome (and the *Sxr* factor) in the mouse[21] had led to the assumption that they would be generally found on the mammalian Y chromosome and

Table 1: Quantitation of Bkm-Related Sequences in the DNA of Various Species

	Area Units/μg	
Source of DNA	pCS316 insert	$(GATA)_5$
E. coli	0	0
Salmon sperm	70	10
Drosophila	>300	137
Mice A/HeJ	>300	225
129/SvPas		
female 1	215	119
female 2	204	111
XX," *Sxr*"	214	71
male	268	172
XY, "*Sxr*"	>300	234
Chimpanzee, female	41	35
Gorilla, female	23	19
Human, female	60	51
Human, male	61	55

be associated with sex determination. However, the search for a restriction enzyme to generate a sex-differentiating pattern on Southern blot hybridization for human Bkm sequences has apparently been unsuccessful[3]. Nevertheless, the regional localizations of Bkm sequences on the X chromosome, which we found, could be related to the presence of genes involved in gonadal differentiation. The Xp21 (minor) and Xq21 (major) localizations of Bkm sequences correspond to regions that may contain genes involved in gonadal development. Female patients with balanced X-autosome translocations which involve an X breakpoint at Xq21 frequently show amenorrhea, hypogonadism, and streak gonads[27,28]. The Xp21 band has not been clearly implicated in sexual differentiation, although deletions of the X from p22 to pter have been associated with male hypogonadism[29,30]. The two autosomal regions, 6q21 and 11q 13-14, have not previously been implicated in sexual differentiation, although the gene for the alpha subunit of chorionic gonadotrophin has been mapped to the q11 to q21 region of 6[31,32].

It is intriguing that both human chromosome 6 and mouse chromosome 17 have regional localizations of Bkm-related sequences. The two chromosomes are partially homologous, since each contains the species-specific Major Histocompatibility Complex (MHC). However, the regional localization is very close to the MHC in the mouse while it is separated by some distance in man (q21 for Bkm, p22 for HLA). This apparent shift in localization from very near the MHC on mouse 17 to the opposite side of the centromere from the MHC on human 6 is also found for mitochondrial superoxide dismutase[33,34] so homology is, in a sense, maintained. Thus, there seems to be conservation of regional localizations of Bkm sequences between mouse and man on the autosomes but not on the Y chromosome.

Characterization of human X-chromosomal, Bkm-containing clones

A library of the human X chromosome in the lambda Charon 21A vector (kindly provided by Dr. Louis Kunkel) was screened with pCS316 for phage that carry Bkm-related sequences. Of about 10,000 plaques screened, 33 were chosen because of their strong hybridization. Nine plaques were purified and shown to contain DNA that was unique in terms of restriction patterns generated by specific enzymes. Five of the clones have been mapped by restriction analysis and the Bkm-homologous fragments identified by Southern analysis with pCS416. We have compared these results with previously published data on Bkm-homologous DNA in *E. radiata*, *D. melanogaster*, and *M. musculus*, which are summarized in Table II. The lengths of Bkm-homologous regions which we have isolated (0.25-1.75 kb) are similar to the lengths of GATA/GACA regions in other organisms. Epplen et al[5,35] prepared detailed restriction maps of Bkm clones from *E. radiata* and *M. musculus* (the largest fragments in the map were about 0.2 kb) and certain restriction sites remained in the GATA/GACA repeat regions[5,35]. For example, *E. radiata* has a GATA/GACA repeat that spans 189 base pairs, with only 2 short, non-repeat regions, but there is an *Alu* I restriction site in the middle of the 80.4% GATA/GACA region[5]. We used a set of enzymes similar to that used by Epplen et al[5,35] but we obtained maps which contained many fewer restriction sites for these enzymes. In addition, none of these restriction sites was found internally in the human Bkm-homologous regions. Thus, although it appears that human Bkm-homologous regions are about the same length as those in other organisms, they may have relatively high concentrations (>80%) of GATA/GACA repeats. We will need to sequence some of our human X-chromosomal Bkm inserts to confirm these tentative conclusions.

Transcription of Bkm-related sequences in man

Previous searches for mRNA transcripts which hybridize to Bkm-related sequences, and which are possibly related to sex determination, have been limited to experiments with adult mouse tissues[5,6,7]. Since we hypothesize that transcripts which relate to sex determination would be expressed at the time of gonadal differentiation, we have studied human fetal tissues. We were able to obtain a variety of tissues from 13 1/2 to 21 week human fetuses aborted by dilation and extraction (the most relevant, earlier material was not available). While no drugs which might alter tissue metabolism are used in this method of performing therapeutic abortions, the fetal remnants are fragmented and many tissues cannot be identified. Thus, we established the sex of the fetuses by analyzing DNA with the human Y chromosomal centromeric probe, Y97[36]. Kidneys, lung, heart, and placenta represent positively identified tissues while several "glands" might have included gonads.

Our initial Northern analyses of these tissues used nick-translated pCS316. This probe contains DNA of unique sequence as well as GATA and GACA repeats, and is analogous to the Elaphe radiata, genomic, Bkm-containing clone used by Epplen et al[5]. We were not able to identify any positive transcripts using this probe, while Epplen et al[5] did not detect sex-specific transcripts when they used GATA and GACA probes[5,7]. Singh et al[6] detected male-specific transcripts in adult mouse liver, using a single-stranded DNA probe of high specific activity from a Drosophila Bkm-related probe, similar to pCS316. We decided, therefore, to prepare a single-stranded RNA probe, since RNA:RNA hybridization is more stable than DNA:DNA hybridization, and very high specific activities

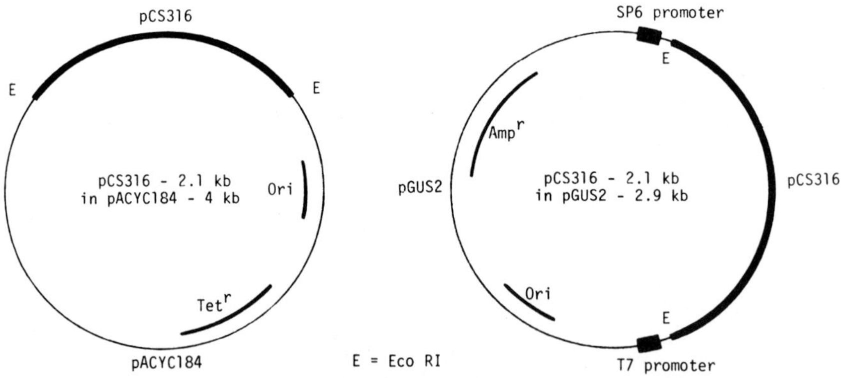

Figure 2: Construction of an anti-sense, Bkm-related clone from pCS316. The Eco RI Drosophila insert from pCS316 was removed and ligated to the polylinker Eco RI site in pGEM™-2 (Promega Biotec). The anti-sense strand was identified by hybridizing single-stranded RNAs, synthesized by SP6 or T7 RNA polymerase, to $(GATA)_5$ at 45°C, in 5xSSPE. The strongly positive signal obtained with the SP6-promoted RNA identified this RNA as the anti-sense RNA.

Table 2: Sequenced Bkm-Homologous Regions

	Elaphe radiate#	Drosophila melanogaster+	Drosophila melanogaster+	Mus musculus*	Mus musculus*
Clone	pErs5	CS3142(8)	CS319	pMC14	M3.1
Length of Clone	2.5 kb	1.5 kb	N/G	2.5 Kb	N/G
Length of region sequenced	2487 bp	545 bp	499 bp	2461 bp	420 bp
Total number of GATA/GACA	152 bp (38 rep)	265 bp (66 rep)	148 bp (37 rep)	544 bp (133 rep)	88 bp (22 rep)
Longest stretch of GATA/GACA	80 bp (20 rep)	52 bp (13 rep)	60 bp (15 rep)	56 bp (14 rep)	32 bp (8 rep)
Span of GATA/GACA repeats	189 bp	421 bp	364 bp	2244 bp	362 bp
% GATA/GACA in span	80.4%	62.7%	40.7%	24.2%	24.3%

KEY: N/G - Information not given
- Reference 5
+ - Reference 6
* - Reference 35
rep = repeats

can be obtained during synthesis of the single-stranded RNA. We prepared the anti-sense construct using a pGEM™-2 (Promega Biotec) (see Figure 2). We used the synthetic oligodeoxynucleotide $(GATA)_5$ to establish the orientation of the insert to the 2 promoters, in order to determine which of them would initiate the transcription of the anti-sense strand.

We found strong hybridization of the single-stranded, anti-sense RNA probe to most of the fetal tissues. The predominant hybridization corresponded to the position of the 18S ribosomal RNA. We do not yet know if this hybridization represents cross-hybridization to 18S RNA. However, it cannot be merely nonspecific hybridization to the large amount of ribosomal RNA, since very little hybridization occured at the position of the 28S rRNA. Preliminary data indicate that the anti-sense Bkm RNA does not hybridize to a cloned fragment of human 18S rRNA. Thus, it seems more likely that we are detecting a mRNA of similar mobility to 18S rRNA. Nonetheless, both this strong hybridization and weaker hybridization to smaller RNAs was roughly equal in males and females, and from 13 1/2 to 21 weeks. Although further characterization of these transcripts is required, none of them seem likely to be related to sex determination.

CONCLUSION

Our results on human Bkm-related sequences emphasize the current lack of evidence for a role in sex determination of GATA/GACA repeats. It is becoming increasingly clear that short simple sequence motifs occur more frequently than can be expected by chance in genomic DNA[37]. Thus, it is unclear whether the GATA/GACA repeats, which are found so widely distributed among animal species, are related by evolutionary descent or have arisen independently[4]. If the repeats have arisen independently, it is less likely that they share a similar function in these diverse organisms. Their frequent location on the heterogenetic chromosome may be indicative of a nonfunctional role - the Y chromosome in mammals is apparently a repository for "junk" DNA[38]. Our finding that higher primates contain fewer Bkm-related sequences than do mice, which are so often used as models for studies of sex determination, might also indicate a nonessential role for many Bkm-related sequences in mammals. However, the higher primates which we studied all had similar amounts of Bkm-related sequences, a result which contrasts with the marked variation in amounts of *Alu* and *Kpn* I repeats found among these higher primates[39]. This conservation of the amount of Bkm-related sequences in the *Hominoidae* might be related to the rate of evolutionary change in simple repeats, as compared to the rate of evolutionary change in the more complex repeats represented by *Alu* and *Kpn* I. Clearly, no role can yet be assigned to Bkm-related sequences in sex determination.

ACKNOWLEDGEMENTS

Portions of the work described in this paper were supported by a Biomedical Research Support Grant (BRSG) and by NIH grant HD20670. We thank Mrs. Diane Voss for her excellent preparation of the manuscript.

REFERENCES

1. Singh, L, Purdom, IF, Jones, KW: Satellite DNA and evolution of sex chromosomes. *Chromosoma*, (Berl.) 59:43-62, 1976.

2. Singh, L, Purdom, IF, Jones, KW: Sex chromosome associated satellite DNA: Evolution and conservation. *Chromosoma*, (Berl.) 79:137-157, 1980.

3. Singh, L, Purdom, IF, Jones, KW: Conserved sex-chromosome-associated nucleotide sequences in eukaryotes. *Cold Spring Harbor Symp Quant Biol*, 45:805-813, 1981.

4. Levinson, G, Marsh, JL, Epplen, JT: Cross-hybridizing snake satellite, *Drosophilia*, and mouse DNA sequences may have arisen independently. *Mol Biol Evol*, 2:494-504, 1985.

5. Epplen, JT, McCarrey, JR, Sutou, S, Ohno, S: Base sequence of a cloned snake W-chromosome DNA fragment and identification of a male-specific putative mRNA in the mouse. *Proc Nat Acad Sci*, *USA*, 79: 3798-3802, 1982.

6. Singh, L, Phillips, C, Jones, KW: The conserved nucleotide sequences of Bkm, which define *Sxr* in the mouse, are transcribed. *Cell*, 36: 111-120, 1984.

7. Schafer, R, Boltz, E, Becker, A, Bartels, F, Epplen, JT: The expression of the evolutionarily conserved GATA/GACA repeats in mouse tissues. Chromosoma, (Berl.) 93:496-501, 1986.

8. Sutcliffe, JG, Milner, RJ, Bloom, FE, Lerner, RA: Common 82-nucleotide sequence unique to brain RNA. Proc Nat Acad Sci USA, 79:4942-4946, 1982.

9. Sutcliffe, JG, Milner, RJ, Gottesfeld, JM, Lerner, RA: Identifier sequences are transcribed specifically in brain. Nature, 308:237-241, 1984.

10. Minarovits, J, Kovacs, Z, Foldes, I: Homologies between a brain-specific identifier (ID) sequence and regions of Harvey murine sarcoma virus and Rous sarcoma virus genomes. FEBS Lett, 174: 208-210, 1984.

11. Owens, GP, Chandhari, N, Hahn, WE: Brain "identifier sequence" is not restricted to brain: Similar abundance in nuclear RNA of other organs. Science, 229:1263-1265, 1985.

12. Shepherd, JCW, McGinnis, W, Carrasci, AE, DeRobertis, EM, Gehring, WJ: Fly and frog homoeo domains show homologies with yeast mating type regulatory proteins. Nature, 310:70-71, 1984.

13. Laughon, A, Scott, MP: Sequence of a Drosophila segmentation gene: Protein structure homology with DNA-binding proteins. Nature, 310:25-31, 1984.

14. McGinnis, W, Garber, RL, Wirz, J, Kuroiwa, A, Gehring, WJ: A homologous protein-coding sequence in Drosophila homeotic genes and its conservation in other metazoans. Cell, 37:403-408, 1984.

15. Carrasco, AE, McGinnis, W, Gehring, WJ, deRobertis, EM: Cloning of a X. laevis gene expressed during early embryogenesis, coding for a peptide region homologous to Drosophila homeotic genes. Cell, 37:409-414, 1984.

16. Rubin, MR, Toth, LE, Patel, MD, D'Eustachio, P, Nguyen-Huu, MC: A mouse homeo box gene is expressed in spermatocytes and embryos. Science, 233:663-667, 1986.

17. Wolgemuth, DJ, Engelmyer, E, Duggal, RN, Gizang-Ginsberg, E, Mutter, GL, Ponzetto, C, Viviano, C, Zakeri, ZF: Isolation of a mouse cDNA clone coding for a developmentally regulated, testis-specific transcript containing homeo box homology. EMBO, 5:1229-1236, 1986.

18. Chandra, HS: Sex determination: An hypothesis based on noncoding DNA. Proc Nat Acad Sci USA, 82:1165-1169, 1985.

19. Wolfe, J, Erickson, RP, Rigby, PWJ, Goodfellow, PN: Cosmid clones derived from both euchromatic and heterochromatic regions of the human Y chromosome. EMBO J, 3:1997-2003, 1984.

20. Thomas, PS, Farquhar, MN: Specific measurement of DNA in nuclei and nucleic acids using diaminobenzoic acid. Anal Biochem, 89:35-44, 1978.

21. Singh, L, Jones, KW: Sex reversal in the mouse (Mus musculus) is caused by a recurrent, nonreciprocal crossover involving the X and an aberrant Y chromosome. Cell, 28:205-216, 1982.

22. Schafer, R, Ali, S, Epplen, JT: The organization of the evoluationarily conserved GATA/GACA repeats in the mouse genome. Chromosoma, (Berl.) 93:502-510, 1986.

23. Arnemann, J, Jakubiczaka, Schmidtke, J, Schafer, R, Epplen, JT: Clustered Bkm DNA sequences on the human Y chromosome. Human Genet, 1986.

24. Kiel-Metzger, K, Erickson, RP: Regional localization of sex-specific Bkm-related sequences on proximal chromosome 17 of mice. Nature, 310:579-581, 1984.

25. Washburn, LL, Eicher, EM: Sex reversal in XY mice caused by dominant mutation on chromosome 17. Nature, 303:338-340, 1983.

26. Kiel-Metzger, K, Warren, G, Wilson, GN, Erickson, RP: Evidence that the human Y chromosome does not contain clustered DNA sequences (Bkm) associated with heterogametic sex determination in other vertebrates. New England J Med, 313:242-245, 1985.

27. Summitt, RL, tipton, RE, Wilroy, Jr, RS, Martens, RP, Whelan, JP: X-autosome translocations: A review. In Sex Differentiation and Chromosomal Abnormalities, ed. Summitt, RL, Bergsma, D. Birth Defects Original Article Series, XIV, 16C, 219-247, 1978.

28. Kaiser, P, Harprecht, W, Steuernagel, P, Daume, E: Long arm deletions of the X chromosome and their symptoms: A new case (6p q24) and a short review of the literature. Clin Genetics, 26:433-439, 1984.

29. Pfeiffer, RA; Observations in a case of an X/Y translocation, t(X;Y) (p22;q11) in a mother and son. Cytogenet Cell Genet, 26:150-157, 1980.

30. Mataxotou, C, Ikkos, D, Papgiotopoulou, P, Alerizaki, M, Marrou, A, Tsenghi, C, Matsaniotis, N: A familial X/Y translocation in a boy with ichthyosis, hypogonadism, and mental retardation. Clin Genet, 24:280-383, 1983.

31. Naylor, SL, Sakaguchi, AY, Chin, WW, Jacobs, J, Shen, LP, Rutter, WJ, Lalley, PA: Chromosomal mapping of mouse hormone genes: somatostatin, calcitonin, thyrotropin alpha subunit, and luteinizing hormone beta subunit. Cytogenet Cell Genet, 37:551, 1984.

32. Robson, EB, Lamm, LU: Report of the committee on the genetic constitution of chromosome 6. Cytogenet Cell Genet, 37:47-70, 1984.

33. Szymura, TC, Wahl, MR, Klein, J: Mouse mitochondrial superoxide dismutase locus is on chromosome 17. Immunogenetics, 14:231-240, 1981.

34. Buckle, VJ, Edwards, JH, Evans, EP, et al: Chromosome maps of man and mouse II. Clin Genet, 26:1-11, 1984.

35. Epplen, JT, Cellini, A, Romero, S, Ohno, S: An attempt to approach the molecular mechanism of primary sex determination: W- and Y-chromosome conserved simple repetitive DNA sequences and their differential expression in RNA. J Exp Zool, 228:205-312, 1982.

36. Wolfe, J, Darling, SM, Erickson, RP, Craig, IW, Buckle, VJ, Rigby, PWJ, Willard, HF, Goodfellow, PN: Isolation and characterization of an alphoid centromeric repeat family from the human Y chromosome. J Mol Biol, 182:477-485, 1985.

37. Tautz, D, Trick, M, Dorer, GA: Cryptic simplicity in DNA is a major source of genetic variation. Nature, 322:652-656, 1986.

38. Erickson, RP: Evolution of Four Human Y Chromosomal Unique Sequences, submitted to J Mol Evol.

39. Hwu, HR, Roberts, JW, Davidson, EH, Britten, RJ: Insertion and/or deletion of many repeated DNA sequences in human and higher ape evolution. Proc Nat Acad Sci, USA, 83:3875-3879, 1986.

LOCALIZATION OF A GENE FOR THE MALE ENHANCED ANTIGEN ON HUMAN AND MOUSE CHROMOSOMES

Yun-Fai Lau

Howard Hughes Medical Institute
Departments of Physiology and Medicine
University of California
San Francisco, California 94143-0724

There is a paucity of studies at the molecular level of the male-specific H-Y antigen despite rapid advances in recombinant DNA technology in recent years. The major problem lies in the immunological definitions of this antigen. The H-Y antigen which was first demonstrated by transplantation experiments[1] can also be detected by two other assays: the cell-mediated cytotoxicity test and the serological test[2,3]. Many aspects of these systems for the detection of the H-Y antigen are reviewed extensively in the present volume and in the literature[4,5,6]. However, no consensus has been reached by various investigators on whether these immunological methods detect the same or different antigens. Furthermore, both transplantation and cell-mediated cytotoxicity tests are restricted by the major histocompatibility complex (MHC), and can only be performed consistently when highly congeneic animals are used. Studies of the serologically defined H-Y antigen are usually hindered by the low titers of the relevant antisera or monoclonal antibodies and by the technically demanding procedures[7,8,9].

Despite the variability in the methods used for detection of H-Y antigen, studies on the serologically defined H-Y antigen from several laboratories have demonstrated the following general properties of the molecules detected by this immunological method[5,10,11]. H-Y antigen is a conserved, male-specific polypeptide which is expressed in all male cells and at high levels in the Sertoli and sperm cells in the testis. The molecular weight of the major molecular species is estimated to be between 18 and 20 kD. Recent studies using antisera of higher titer have revealed reactive molecules with molecular weights of 20, 50, 66 and 200 kD[12] (Goldberg, this volume). The nature of these different molecular forms of the antigen is not clear. However, the possibility that they are different entities has been suggested[12]. The antigenicity of most anti-H-Y antibodies may be associated with the carbohydrate moiety of the antigen[13,14]. The H-Y antigen has been postulated to be an inducer of testicular differentiation during mammalian embryogenesis[15,16], and we have, for the past few years, been trying to identify the structural gene for the serologically defined H-Y antigen and to characterize this antigen in molecular terms by use of recombinant DNA methodologies. Recently, we have isolated a candidate for this structural gene from a prokaryotic expression (lambda gt11) mouse testicular cDNA library, using an immunological screening procedure with an anti-H-Y serum provided by Dr. Ellen Goldberg of the University of New Mexico. Because of the difficulties in the identification of the serologically defined H-Y antigen, we have termed this gene the male-enhanced antigen (MEA) gene. Molecular characterization of the MEA gene has revealed certain properties which are similar to those of the serologically defined H-Y antigen. These properties include genetic and evolutionary conservation in all mammalian species at both the genomic and the transcriptional level, male-enhanced expression, and Sertoli cell-specific and testis-specific expression (Lau and Goldberg, in preparation). Further studies have reveled that two other related genes are closely linked to the MEA gene in both the human and the mouse (Lau, unpublished data). These observations suggest the possibility of a family of MEA-related genes in both species. Even though the exact correlation of the MEA gene with the gene for the

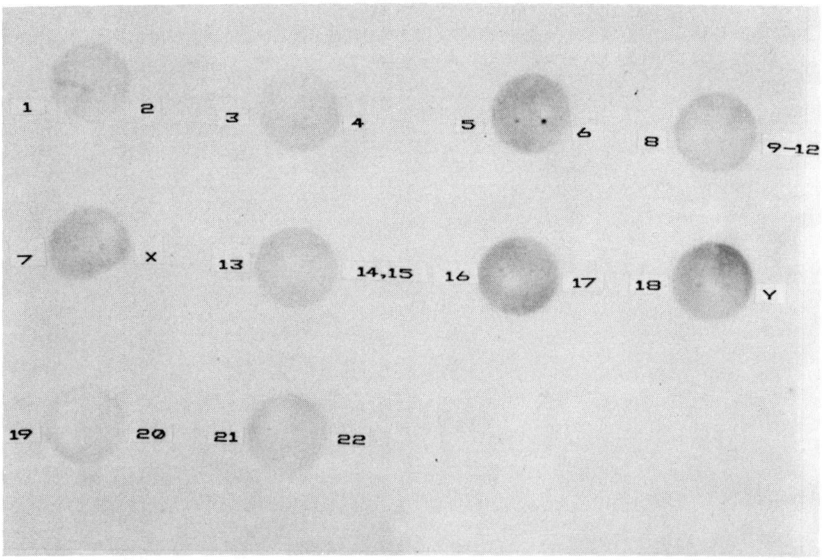

Figure 1. Localization of the MEA gene on human chromosome 6. Human chromosomes were sorted directly onto nitrocellulose filter papers with a dual-laser FACS VI and processed as previously described[17]. The chromosome dots were hybridized with a ^{32}P-labelled human cDNA probe by the established Southern hybridization procedure. Positive hybridization was observed in the autoradiogram over the dot which corresponded to chromosome 6, thereby permitting assignation of the MEA gene to this chromosome.

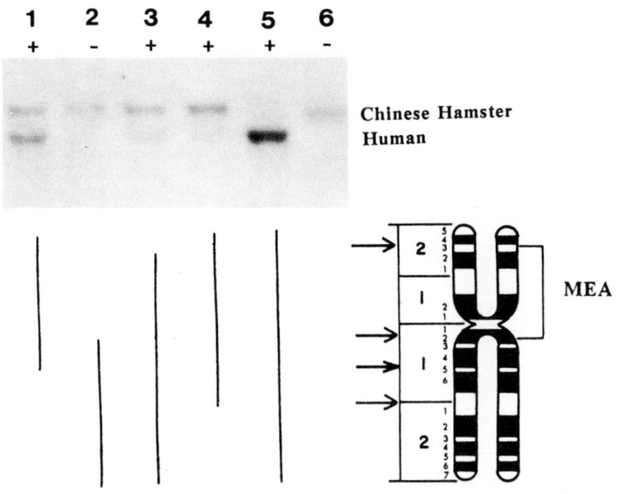

Figure 2. Assignation of the MEA gene to a region of chromosome6. DNA from 4 human-Chinese hamster hybrid cell lines (lane 1-4), which contained different portions of the human chromosome 6 as represented by the lines under each lane and the idiogram on the lower right, were digested with Eco Rl and analysed by Southern hybridization with human cDNA as probe. Lanes 5 and 6 contained human and Chinese hamster DNA, respectively, as controls. The Chinese hamster- and human-specific bands were labelled (top right). Somatic cell hybrids (1, 3, and 4), containing the portion of human chromosome 6 between p^{23}-q^{12}, hybridized positively with the radioactive probe, thereby permitting assignation of the MEA gene to this subregion of chromosome 6.

serologically defined H-Y antigen requires more extensive investigations, the genetic and evolutionary conservation and male-enhanced expression of the family of MEA genes indicate that this family must serve important functions in the mammalian testis.

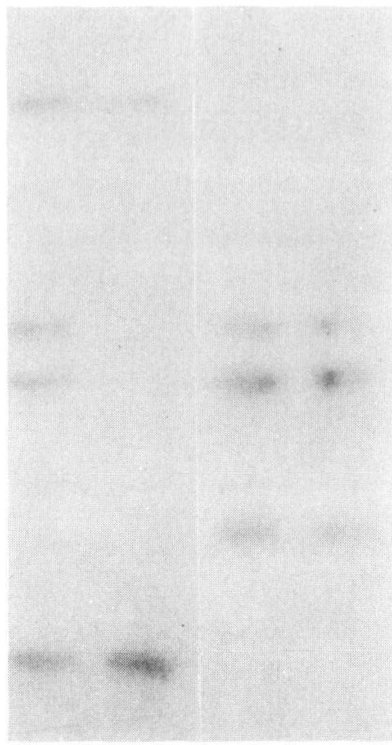

Figure 3. Assignation of the MEA gene to specific mouse chromosomes. DNAs from a mouse-Chinese hamster somatic cell hybrid which contained mouse chromosome 17 and part of chromosome 18 (lane 1), from Chinese hamster cells (lane 2), and from male (lane 3) and female (lane 4) mice were digested with Bgl II and analyzed by Southern hybridization with the mouse cDNA as probe. Two of the mouse-specific bands (+) were present in the DNA from the somatic cell hybrid, indicating that this cell line contains the mouse MEA gene. One mouse-specific band (-) was absent in this cell line.

Southern blot analysis of DNA from males and females suggests that the MEA gene is located on an autosome. Using human cDNA as a probe and the chromosome dot hybridization technique[17], we have mapped the MEA gene to human chromosome 6 (Figure 1). To define more exactly the location of the MEA gene on chromosome 6, we have used Southern blot analysis of DNAs isolated from several somatic cell hybrids which harbor different portions of human chromosome 6 in a Chinese hamster background[18,19]. The results from this analysis indicate that the MEA gene is located between bands p^{23}-q^{12} on this chromosome (Figure 2). The mapping of the MEA gene on mouse chromosomes is somewhat complicated. Preliminary Southern blot analyses of DNA obtained from a somatic hybrid which contained mainly mouse chromosome 17 in a Chinese hamster background[20] suggest that most of the MEA-specific DNA sequences are located on this chromosome in the mouse genome (Figure 3). Other studies show that the restriction enzyme patterns of MEA sequences on this chromosome correspond to those of an MEA functional gene isolated from a mouse cosmid library. The chromosomal location and the nature of other related sequences are unknown. Since the human chromosome 6 and mouse chromosome 17 contain similar genetic loci, for example as the major histocompatibility complex, the mapping of the MEA gene to the respective homologous chromosome in both species further confirms the genetic and evolutionary conservation of the family of MEA genes.

The autosomal locations of the MEA gene and its male-enhanced expression present a very complex problem in terms of the regulation of its expression. Presumably, the Y chromosome may, somehow, contain a regulatory gene whose product either directly or indirectly enhances the expression of the MEA gene. Molecular characterization of the promoter sequences of the MEA gene and of factors which bind to these sequences should provide some insights into the mechanisms which regulate its tissue-specific expresion. The probable relationship between the MEA regulatory gene and the testis-determining factor (TDF) gene on the human Y chromosome (Human Gene Mapping Workshop 8, 1985) or the similar locus, Tdy, on the mouse Y chromosome[21], is not clear. However, these loci may be not mutually exclusive. The provisional assignment of the MEA gene to chromosome 17 in the mouse is an particularly interesting observation. Many class I histocompatibility antigens located on chromosome17 are expressed on preimplantation embryos, and presumably play a role in the control of cellular interactions during embryogenesis[22]. The expression of the serologically defined H-Y antigen on preimplantation embryos has been demonstrated[23]. Kralova and Demant[24] showed that a locus, Hye, which regulates the level of expression of H-Y antigen, is located on chromosome 17, somewhere near the T complex[25]. Erickson[22] suggested that the Hye locus may encode a galactosyltransferase. Washburn and Eicher[26] also mapped a primary, autosomal, sex-determining locus, the T-associated sex-reversal (Tas,) locus, on chromosome 17. Mutations at this locus result in XY mice whose gonads develop into either ovaries or ovotestis. Although the relationship between the family of MEA genes and the Hye and Tas loci is uncertain, the provisional mapping of the MEA gene on chromosome 17, if confirmed, places the family of MEA genes on the portion of the mouse genome that is important for embryogenesis and sex determination. The exact functions of the family of MEA genes can only be elucidated by further vigorous molecular studies of this family of genes at the molecular level.

ACKNOWLEDGEMENTS

I thank Drs. K.-H. Grzeschik and F. Ruddle for providing DNA from the somatic cell hybrids, Dr. Y. W. Kan for support and encouragement, and Drs. R.V. Lebo, K. Chan and N. Zoghbi for technical assistance. Y.-F. Lau is an Associate Investigator of the Howard Hughes Medical Institute.

REFERENCES

1. E. J. Eichwald and C. R. Silmser, Untitled communication. Transplant Bull 2:148-149 (1955).

2. E. Goldberg, E. A. Boyse, D. Bennett, M. Scheid, and E. A. Carswell, Serological demonstration of H-Y (male) antigen on mouse sperm. Nature 232:478-480 (1971).

3. E. H. Goldberg, F. Shen, and S. Tokuda, Detection of H-Y (male) antigen on mouse lymph node cells by the cell to cell cytotoxicity test. Transplantation 15:334-336 (1973).

4. E. Simpson, The role of H-Y as a minor transplantation antigen. Immunol Today 3:97-106 (1982).

5. S. S. Wachtel, H-Y Antigen and the Biology of Sex Determination. Grune and Stratton, New York (1983).

6. P. W. Andrew, The male-specific antigen (H-Y) and sexual differentiation. In Genetic Analysis of the Cell Surface, P. Goodfellow, ed., Chapman and Hall, London. pp 159-190 (1984).

7. M. T. Zenzes and T. E. Reed, Variability in serologically detected male antigen titer and some resulting problems: a critical review. Hum Genet 66:103-109 (1984).

8. M. T. Zenzes and T. E. Reed, Untitled reply. Hum Genet 66:113-114 (1984).

9. A. Mayerova, U. Muller, U. Wiberg, U. Wolf, and M. Fraccaro, Variability in the amount of serologically detectable H-Y antigen. Hum Genet 66:110-112 (1984).

10. Y. Nagai, H. Iwata, D. D. Stapleton, R. C. Smith, and S. Ohno, Testis-organising H-Y antigen of man may lose its receptor-binding activity while retaining antigenic determinants. In Testicular Development, Structure and Function, A. Steinberger and E. Steinberger, eds. Raven Press, New York, pp 41-47 (1980).

11. J. L. Hall, Y. Bushkin, and S. S. Wachtel, Immunoprecipitation of human H-Y antigen. Hum Genet 58:34-36 (1981).

12. M. P. Bradley, B. F. Heslop, A biochemical and immunological approach to the identification of H-Y antigenic proteins secreted from Daudi cells. Hum Genet 71:117-121 (1985).

13. M. Shapiro and R. P. Erickson, Evidence that the serological determinant of H-Y antigen is carbohydrate. Nature 290:503-505 (1981).

14. M. Shapiro and E. H. Goldberg, Analysis of a serological determinant of H-Y antigen: evidence for carbohydrate specificity using an H-Y specific monoclonal antibody. J Immunogenet 11:209-218 (1984).

15. S. S. Wachtel, S. Ohno, G. C. Koo, and E. A. Boyse, Possible role for H-Y antigen in the primary determination of sex. Nature 257:233-236 (1975).

16. F. P. Haseltine and S. Ohno, Mechanisms of gonadal differentiation. Science 211:1272-1278 (1981).

17. R. V. Lebo, F. Gorin, R. J. Fletterick, F.-T. Kao, M.-C. Cheung, B. D. Bruce, and Y. W. Kan, High-resolution chromosome sorting and DNA spot-blot analysis assign McArdle's syndrome to chromosome 11. Science 225:57-59 (1984).

18. K.-H. Grzeschik, Assignment of a structural gene for glucuronidase to human chromosome C7. Somatic Cell Genet 2:401-410 (1976).

19. K.-H. Grzeschik, A. Mayerova, K. Bender, and R. Johannsmann, Regional submapping of human chromosome 6 by the use of human-mouse somatic cell hybrids. Cytogenet Cell Genet 22:412-414 (1978).

20. P. D'Eustachio and F. H. Ruddle, Somatic cell genetics and gene families. Science 220:919-924 (1983).

21. T. H. Roderick and M. T. Davisson, Linkage Map of the Mouse, The Jackson Laboratory, Bar Harbor (1984).

22. R. P. Erickson, Genetics of the cell surface of the preimplantation embryo:studies on antigens determined by chromosome 17 in the mouse. In Genetics Analysis of the Cell Surface, P. Goodfellow, ed., Chapman and Hall, London, pp 143-157 (1984).

23. C. J. Krco and E. H. Goldberg, Detection of H-Y (male) antigen on 8-cell mouse embryos. Science 193: 1134-1135 (1976).

24. J. Kralova and P. Demant, Expression of the H-Y antigen in thymus cells and skin: differential genetic control linked to K end of H-2. Immunogenetics 3:585-594 (1976).

25. J. Kralova and A. Lengerova, H-Y antigen: genetic control of the expression as detected by host-versus-graft popliteal lymph node enlargement assay maps between the T and H-2 complexes. J Immunogenet 6:429-438 (1979).

26. L. L. Washburn and E. M. Eicher, Sex reversal in XY mice caused by dominant mutation on chromosome 17. Nature 303:338-340 (1983).

ABBREVIATIONS

BFO	Bovine fetal ovarian (cells)
Bkm	Banded Krait minor
CAM	Cell adhesion molecule
CMC	Cell-mediated cytotoxicity
CTL	Cytotoxic T lymphocytes
DTH	Delayed type hypersensitivity
FCS	Fetal calf serum
MEA	Male enhanced antigen
MHC	Major histocompatibility complex
PLN	Popliteal lymph node
PrA-SRBC	Protein A - sheep red blood cell

Other abbreviations pertain to actual or hypothetical genotypes and are not included. They are explained individually within each chapter.

CONTRIBUTORS

Dr. David Crews
Department of Zoology
University of Texas
Austin, Texas 78712

Dr. Albert de la Chapelle
Department of Medical Genetics
University of Helsinki
00290 Helsinki 29, Finland

Dr. Ernst J. Eichwald
University of Utah Medical Center
50 North Medical Drive
Salt Lake City, Utah 84132

Dr. Robert P. Erickson
Department of Human Genetics, Box 015
4708 Medical Sciences II
The University of Michigan
Ann Arbor, Michigan 48109-0010

Dr. Ellen Goldberg
Department of Microbiology
School of Medicine
The University of New Mexico
Albuquerque, New Mexico 87131

Dr. Peter N. Godfellow
Laboratory of Human Molecular Genetics
Imperial Cancer Research Fund
Lincoln's Inn Fields
London WC2 3 PX, England

Dr. Gloria C. Koo
Department of Immunology
Merck, Sharp, and Dohme Research Laboratory
Rm 115, Bldg. ADW
Rathway, New Jersey 07065

Dr. Yun-Fai Lau
Howard Hughes Medical Institute
Department of Physiology and Medicine
University of California
San Francisco, California 94143-0724

Dr. Paul G. McDonough
School of Medicine
Department of Obstetrics and Gynecology
Human Genetics Institute
Medical College of Georgia
Augusta, Georgia 30912

Dr. Anne McLaren
MRC Mammalian Development Unit
Wolfson House, 4 Stephenson Way
London NW 1 2HE, England

Dr. Claude Nagamine
Department of Immunochemistry
Sloan-Kettering Institute for Cancer Research
1275 York Avenue
New York, New York 10021

Dr. Susumo Ohno
Department of Biology, City of Hope National Medical Center
1500 East Duarte Road
Duarte, California 91010

Dr. David Page
The Whitehead Institute
9 Cambridge Center
Cambridge, Massachusetts 02142

Dr. Roger Short
Department of Anat and Physiol
Monash University
Clayton, Victoria 3168, Australia

Dr. Kirby D. Smith
933 Traylor Building
Johns Hopkins University School of Med.
720 Rutland Avenue
Baltimore, Maryland 21205

Dr. Jonathan Wolfe
The Galton Laboratory
Department of Genetics and Biometryrch Group
University College, Wolfson House
4 Stephenson Way, London NW1 2HE
England

INDEX